HIGH-RISE OFFICE BUILDING

高迪国际出版有限公司　编

中高层办公建筑

李青云　译

大连理工大学出版社
Dalian University of Technology Press

图书在版编目(CIP)数据

中高层办公建筑：英汉对照 / 高迪国际出版有限公司编；李青云译. — 大连：大连理工大学出版社，2012.5

ISBN 978-7-5611-6840-0

Ⅰ.①中… Ⅱ.①高… ②李… Ⅲ.①办公建筑—建筑设计—作品集—世界 Ⅳ.①TU243

中国版本图书馆CIP数据核字（2012）第061165号

出版发行：大连理工大学出版社
（地址：大连市软件园路80号 邮编：116023）
印　　刷：利丰雅高印刷（深圳）有限公司
幅面尺寸：240mm×320mm
印　　张：21
插　　页：4
出版时间：2012年5月第1版
印刷时间：2012年5月第1次印刷
责任编辑：裘美倩
责任校对：王秀媛
装帧设计：屈舒丽

ISBN 978-7-5611-6840-0
定　　价：328.00元

电 话：0411-84708842
传 真：0411-84701466
邮 购：0411-84703636
E-mail: designbook@yahoo.cn
URL: http://www.dutp.cn

如有质量问题请联系出版中心：（0411）84709043 84709246

HIGH-RISE
OFFICE
BUILDING

RESPONSIBLE PLANNING

THAT'S WHAT HUMANITY ASKS FOR.

We architects have a great responsibility for each project we plan, because all our work will have a positive or negative social and economic impact on the community's future.

When we design office blocks we need to make an in-depth analysis of the context in which they will be built. Studying the territory carefully makes it possible to plan the principles of bioclimatics and orient the building in the ideal position to allow the sun's rays to enter in winter and be shielded from them in summer. In this way maximum comfort is ensured throughout the year, even without using air conditioning systems.

An analysis of a building's shape should have the purpose of saving energy and enhancing the social nature of the construction. The idea behind the architectural project should be conceived bearing in mind environmental sustainability, behavioural ethics and a good social impact. Architecture is not just sculptural shape, but also function, communication and effectiveness, because it is in the empty spaces inside this sculpture that we will live our working lives. The right shape, combined with the use of materials that can place at our disposal technological and aesthetic qualities, help to enhance a building's performance and achieve energy savings of about 40% without employing machines. We therefore need to unite the principles of bio-architecture with integrated planning so as to guarantee wellbeing in quality buildings that perform well and are in perfect harmony with the environment.

To tread the path of building offices that consume less energy is a question of design ethics. Nowadays we have the instruments for improving the quality of the air, saving energy and reducing emissions of CO_2. We have materials that insulate and protect buildings naturally and paints based on nanotechnology that stop smog absorption.

I believe that the boundary between work environment, rest place and space in which to socialise will in the near future be less and less obvious. Operating spaces will feature simple informal styling so that workers can improve their lot and interact with each other. These are flexible work areas that also relate with the greenery outdoors and can host the most sophisticated technologies, without ostentation, using intelligent illumination systems to communicate with the outdoor space.

The furniture will have a "Natural Wellness" effect due to the use of natural antibacterial paints, and the rooms will be made more attractive and comfortable with the use of organically produced natural fibres such as cotton, linen and hemp.

The fusion of interior design and corporate image expresses the company's identity more effectively and develops a sense of belonging in employees. In a society that is continuously evolving, but with buildings that will distinguish the territory for many years, the architect must be an ethical person who pays careful attention to materials, forecasting their aging and adopting those of excellent quality for cladding and coating, in addition to considering aesthetic solutions that will facilitate updating technological equipment over time.

MARIO MAZZER

认真负责的规划：人类的诉求

建筑师对每个规划项目都有着高度的责任感，因为建筑师所做的大大小小的事情会给项目所涉及的社区的未来带来社会方面或者经济方面的影响，这些影响或好或坏。

在设计办公大楼的时候，我们需要对项目所在的背景环境作深入的分析。只有对楼盘所在地域进行认真的研究，才有可能根据当地的生物气候形态规则来做整体规划，例如，让建筑朝向最为理想，冬季阳光能充分照进大楼，而夏季则尽量屏蔽阳光。在这样的大楼内，即使常年不用空调系统，也能最大程度地保持舒适状态。

我们还需要认真研究将大楼设计成什么样的形状以达到既节约能源又能加强大楼社会功能的目的。建筑所传递的理念应该充分考虑环境的可持续性、行为道德标准以及良好的社会声誉。建筑绝不只是展示其雕塑化的外形，还应展示其具有的功能及其与周围环境的互动和成效。适宜的外形与具有科技含量并符合审美标准的建材搭配使用，可大大增强建筑的性能，减少使用机械，可节能 40% 左右。因此我们需要将绿色建筑理念融入整体规划中，以确保建筑质量高、内部环境舒适、性能良好，并与周围环境相协调。

探索建造节能办公室实际是设计伦理问题。当前科技的发展使我们拥有了改善空气质量、节能、降低二氧化碳排放量的各种设备，我们还拥有各种绝缘和保护建筑免受自然毁坏的材料以及基于纳米技术能阻挡烟雾吸附的油漆。

我坚信在不久的将来，办公环境、休息场所以及社交场所的界线将越来越不明显。办公空间将营造出简单、随意的氛围。员工可以根据自己的需要改善工作区，不同区域之间还能够互动。这些布局灵活的办公空间可以装载最精良的技术产品，并使用智能照明系统与户外空间相通，使工作区域与绿化良好的户外空间浑然一体。

家具设计由于使用了天然抗菌油漆而具有自然、健康的效果。室内装饰由于使用了如棉质、亚麻和大麻纤维等有机产品而更加舒适和有吸引力。

将办公大楼的室内设计与公司形象完美融合可以更好地表达公司特性，同时也能培养员工的归属感。社会在不停变化着，而独特存在的建筑工程却要维系多年，这就要求建筑师必须成为具有道德感的人，既要高度关注所使用的建筑材料，能预测建筑材料的使用年限，采用优质的外墙和立面材料；同时也要考虑到随着时间的推移，更新技术设备时的审美解决方案。

MARIO MAZZER

CONTENTS

008 Landmark East Hong Kong
016 Shanghai Huawei Technologies Corporate Campus
028 TT Project
042 Sequana Tower / Bouygues Telecom Headquarters
054 INTERFACE Office Building
064 Fukoku Tower
074 Sanlitun SOHO
084 U15 Office Building
094 IT–Fornebu Portal Building
102 International Management Institute(IMI)
114 Homes Headquarters
126 Man Air
134 55 Park Street Laboratory
142 Hungarian Autoklub Headquarters
152 Elbek & Vejrup
160 Dogan Media Center Ankara
168 Wilkie D. Ferguson Jr. United States Courthouse

AIC Automotive Intelligence Center 178
555 Mission Street 190
CSET–Center for Sustainable Energy Technologies 200
Banco Real Santander Headquarters & WTorre Plaza Mixed-Use 210
UNIQA Vital Business Center Budapest 224
1100 First Street 232
331 Foothill Road Office Building 238
BWT Offices 246
Eastern Harbor International Tower 252
Oficina Vidre Negre 260
Breathing Factory 274
SOLON SE 282
Rivergate 296
The Water Center 308
Torre Arrau 320
INDEX 330

Landmark East Hong Kong

LOCATION\Hong Kong, China

Landmark East is a Grade-A high-rise, two-tower office development that rises from a dense urban site in the Kwun Tong district of Hong Kong, China. The slender rectilinear slabs that form the tower volumes are derived from the long narrow plot. The towers respond to the site, rising above the adjacent buildings in a slender, elegantly moving motion reminiscent of bamboo swaying gently in the wind. The idea was to differentiate the buildings from the static neighborhood in creating a dynamic form that clearly defines the towers from their surroundings.

The rectangular floor plates of both towers provide efficiency and allow side cores to the north. This in turn orientates the office floors to the south, taking advantage of the privileged views of Victoria Harbor and Hong Kong Island beyond. The orthogonal tower forms have been segmented in section creating a series of interlocking parallelogram slabs. This arrangement emphasizes the sensation of movement and conveys lightness to the towers, forming a strong, dramatic, unified identity for the development.

Efficient floor plans are therefore retained whilst providing dynamism to what would otherwise be a regular box derived from a rectangular plan. Where the forms interlock additional corner offices are created taking advantage of the valuable harbor and island views. The car parking and loading areas are restricted to the basement of both towers and a podium structure at the base of tower-two. As a result, the remainder of the site is left open to form an elegant and beautifully landscaped public plaza with sculptures gardens and cafes that provide valuable open space for tenant and local residents use.

The reinforced concrete structure of the towers utilizes high strength concrete to reduce core wall and column area. Post-tensioned band beams were employed to maximize clear ceiling heights of 4.2 meters in the office floors. The building facades consist of a fully unitized aluminum framed curtain wall system. The floor to ceiling glazing is double low-e coated insulated glass allowing maximum light transmission while reducing thermal transmission through the façade.

A brisole system has been applied to the facades, which articulates the building form and provides solar shading. Horizontal fins to the south façade provide high-level sun shading protection; vertical fins on the narrow east and west elevations alleviate low-level horizontal shading protection. An energy efficient VAV system supplies air-conditioning to the project and rainwater is harvested for use in the landscape irrigation system. These and other sustainable components within the design have assisted in obtaining a Provisional Platinum BEAM award.

ARCHITECT
Bernardo Fort Brescia
FIRM
Arquitectonica
AREA
110000m^2
PHOTOGRAPHER
Amral Imran, Rogan Coles

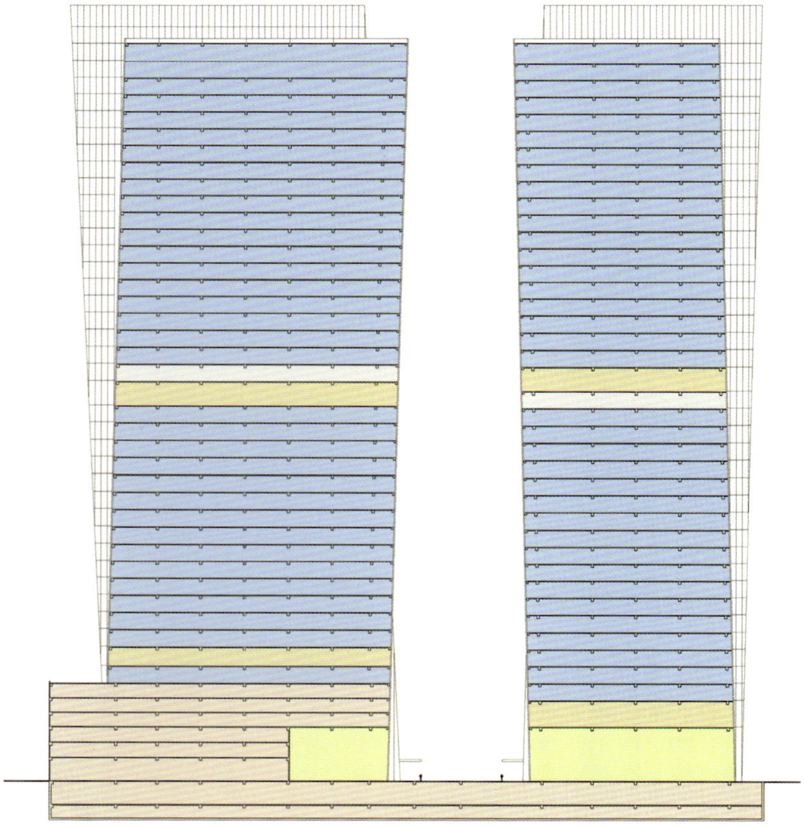

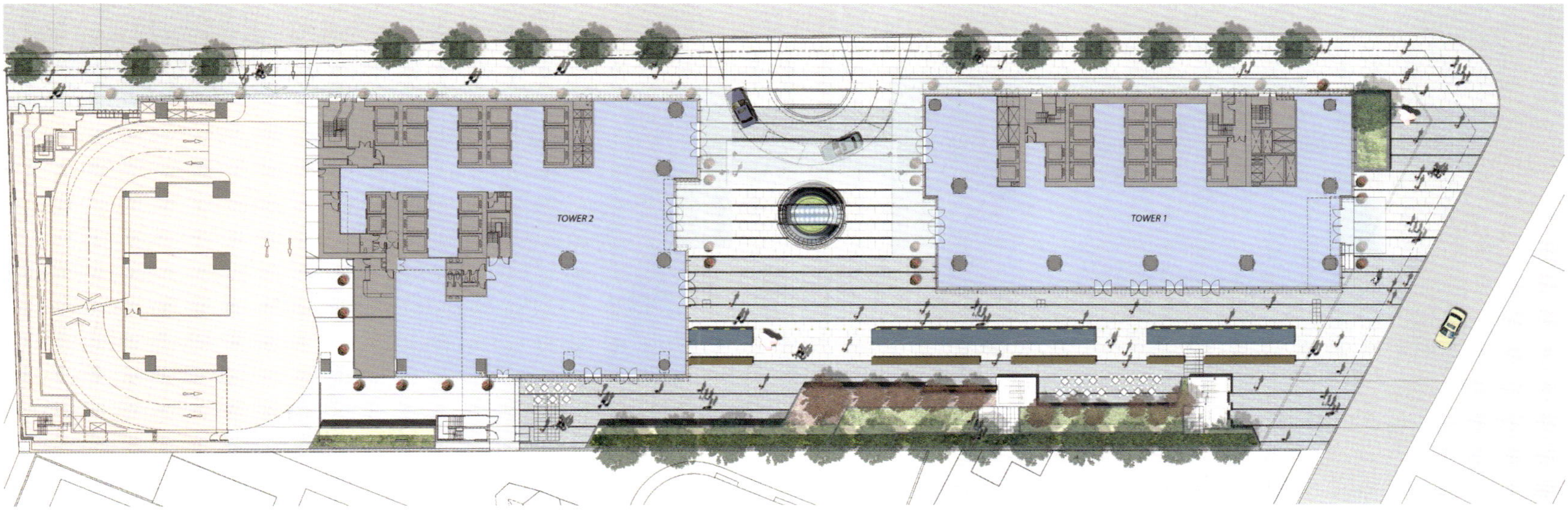

在香港高速发展的观塘中央商务区耸立着一个优质高档的双塔式办公复合建筑——城东志。在一个狭长的空间中，两个简洁的直角板层塔楼形成相互关联的塔体。塔楼本身与该址的特色互相呼应，纤细、优美的动态构图使高楼本身从周围建筑中脱颖而出，就像竹子在风中轻轻摇曳。该建筑的设计理念就是让自己超然于周围呈静态的商业区建筑，而营造一种动感。

直线形楼面提高了利用效率，同时也使大楼侧面的核心区域朝向北方。这种设计使办公楼层朝南，可以饱览维多利亚港湾和香港岛上的风光。两个直角形的分离的塔楼形成相互关联的平面。这样的设计在加强了建筑动感的同时，也使充足的光线穿透进来。这些都为该区的开发形成一个有力、巨大、统一的形象。

各楼层经过特殊高效的设计增加了动感，在所形成的额外的拐角办公室里可以观赏海港和海岛景色。停车场和卸货区设置在双塔的地下室以及二号塔楼地下室的平台结构中。因此，项目地基的其余部分就成了设计精美、景观怡人的开放空间，其中有景致优美的公共广场，广场上设有雕塑园和咖啡厅，方便了租户和当地的住户。

塔体本身的钢筋混凝土结构充分利用了高强度的混凝土，以减少框架和柱子的面积。通过采用高性能混凝土浇筑的带状横梁，办公楼层的室内净高达到了 4.2 米。整体由铝框幕墙系统组成的立面，安装密闭的双层低辐射隔热隔声的玻璃幕墙组成先进的系统，既保证了充足的自然光的入射，同时也降低了热传输功能。

立面应用了 Brisole 系统，既有力地表达了建筑形式，同时也提供了遮阳设施。其中南部外墙水平安装的遮阳叶片提供了较强的遮阴功能，而在狭窄的东西外墙上的垂直的遮光叶片缓和了低级别的水平遮阳防护。空调系统在保护住户的同时提供最佳的能源效率，塔内还设计有收集雨水灌溉景观植物的灌溉系统。设计中上述种种措施以及其他的可持续发展因素使得城东志在 BEAM 评级制度下获得 “白金” 评级。

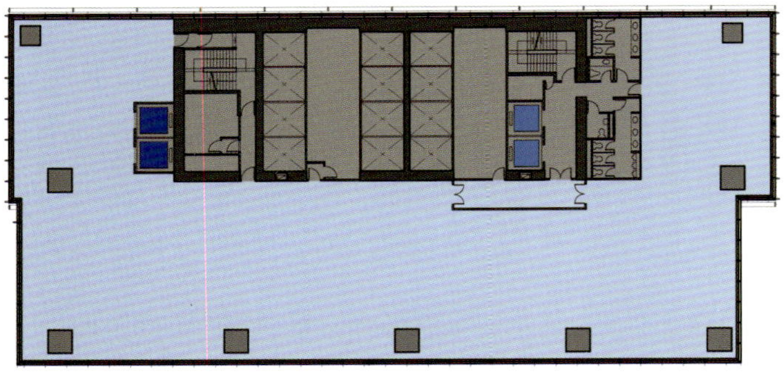

LOW ZONE L2 – L4

LOW ZONE L2 – L12

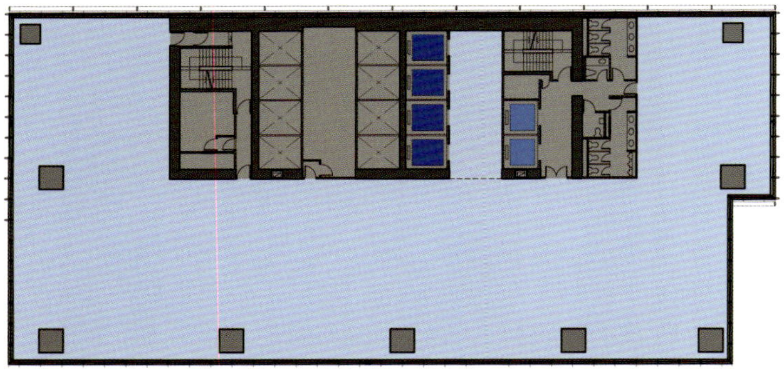

LOW ZONE L5 – L13

MID ZONE L13 – L15

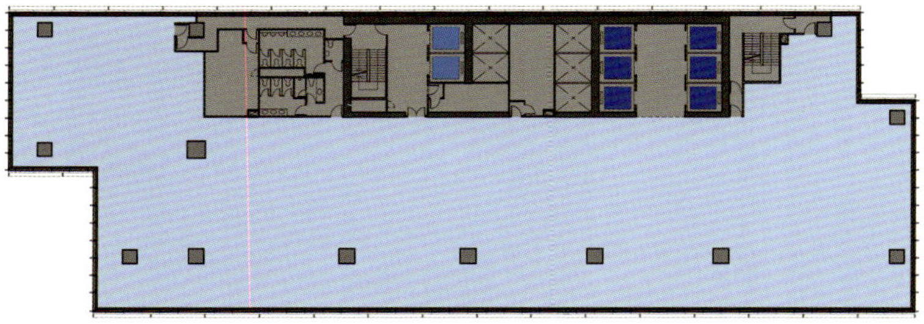

MID ZONE L18 – L25

MID ZONE L14 – L18, L21 – L28

HIGH ZONE L29 – L39

HIGH ZONE L26 – L36

Shanghai Huawei Technologies Corporate Campus

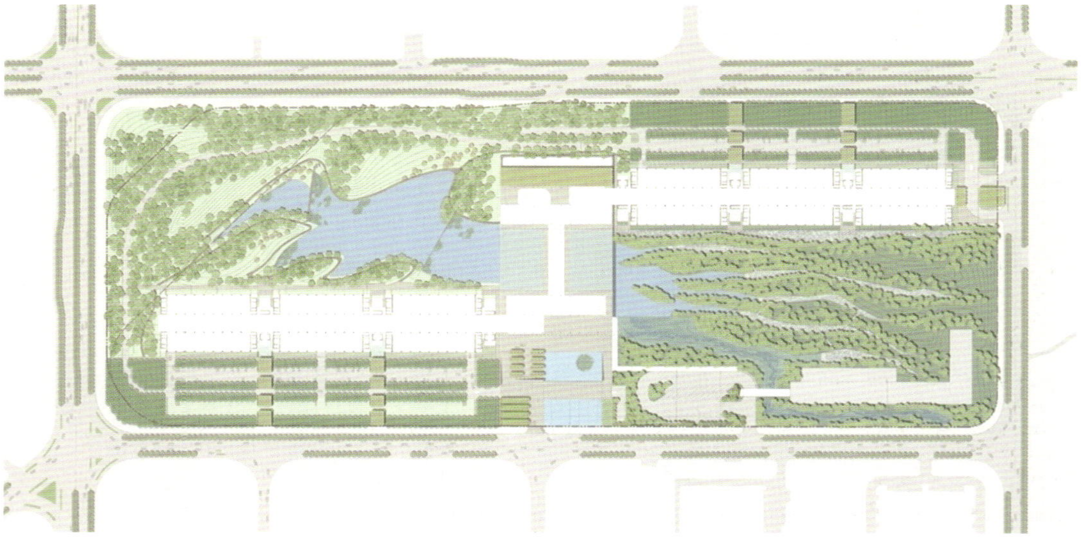

ARCHITECT
SOM
AREA
300000 m^2
PHOTOGRAPHER
SOM | © Tim Griffith

LOCATION\Shanghai, China

The campus is easily accessible from various modes of transportation and a future subway line extension will also connect the campus to downtown Shanghai and beyond.

With a total building area of over 300,000m² of Research and Development space including ancillary areas, the site plan takes advantage of the relatively long and narrow dimensions of the 275,000m² site to create a low-rise workplace that engages a new landscape and capitalizes on the broad south orientation. To reduce the overall length of the buildings, the site plan organizes the Research and Development program into two linear wings of linked five-storey buildings, organized around atria, and attached to a central administrative, meeting, and amenity hub.

Planned to accommodate 7,500 employees, the campus is dedicated to advancing next-generation mobile technology. The interiors are divided into open-office work plans and closed laboratories. Though separate, the various departments are linked through continuous sky-lit galleries. To maximize flexibility and anticipate future change in research and technology, the floor plate modules are designed as a series of 18-by-9-meter structural bays. This design amplifies interior exposure to natural light and exterior views.

Secure building entrances along the gallery concourse provide access to elevator and stairs to the upper levels of R&D office modules that are connected by continuous access balconies and bridges along the gallery. The two R&D buildings are offset on the site to define a central arrival court for the facility. Surrounding the raised central court are non-R&D administrative offices, product display space, conference and training rooms. The staff cafeteria and specialized research facilities are located under the central court. multiple parking levels below each of the R&D buildings accommodate parking for 3,000 vehicles.

The placement of the Research and Development buildings on the site allows for a continuous landscape to traverse the length of the site. In order to mitigate the high water table and deal with the threat of site flooding, the landscape is sculptured into a series of wetlands that form rainwater retention basins and runnels for bio-filtering. These lushly native planted gardens and water features visually complement the office buildings and provide outdoor gathering spaces and breakout areas for the employees.

The building structural systems are efficient and focused on component innovation. The use of a tapering profile of the typical post tensioned (PT) framing led to greater efficiency and less material, saving significant material for the framing system. The repetitive nature of the architectural and structural design of the R&D office modules simplified construction and led to faster, more efficient construction with less material waste, particularly with the re-use of formwork. The compression-only piles used pre-tensioned spun high strength concrete (PHC) and were detailed as hollow pipes with a steel shoe at the bearing end provided saving in concrete material. Recycled material and cement replacements were specified to conserve resources.

Envisioned as a state of the art high performance office environment, the exterior curtain wall for the Research and Development campus is an externally ventilated, double wall of a single and insulated non-reflective low-E exterior glazing system with aluminum panels. Set in the cavity of the double wall are motorized louvers that are connected to the automatic sensor controls to modulate and balance the quality of light into the predominately computer based workspaces. Glass was carefully selected to balance thermal comfort with views and glare reduction. The windows are operable for natural ventilation, smoke exhaust, and cleaning. The double wall reduces thermal loads on building by creating a thermal barrier next to the inner light of glass.

The 18 meters wide galleria provides abundant daylighting and reduces lighting loads to the research functions lining the atrium-like space, but also serves as the gathering place and pathway to transverse the project. The gallerias are passively cooled / heated from office exhaust. Water gardens and bamboo groves create a soft, naturalistic setting for the spine of the complex. Enclosing MEP functions, providing lateral stability, and incorporating sun protection for the exterior glazing, precast concrete roof elements at the R&D buildings unify the multiple buildings into a complex with a consistent and lasting identity.

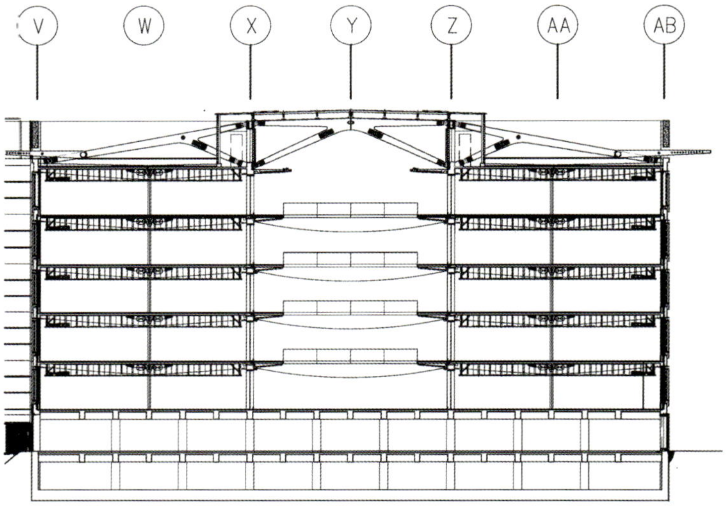
V
W
X
Y
Z
AA
AB

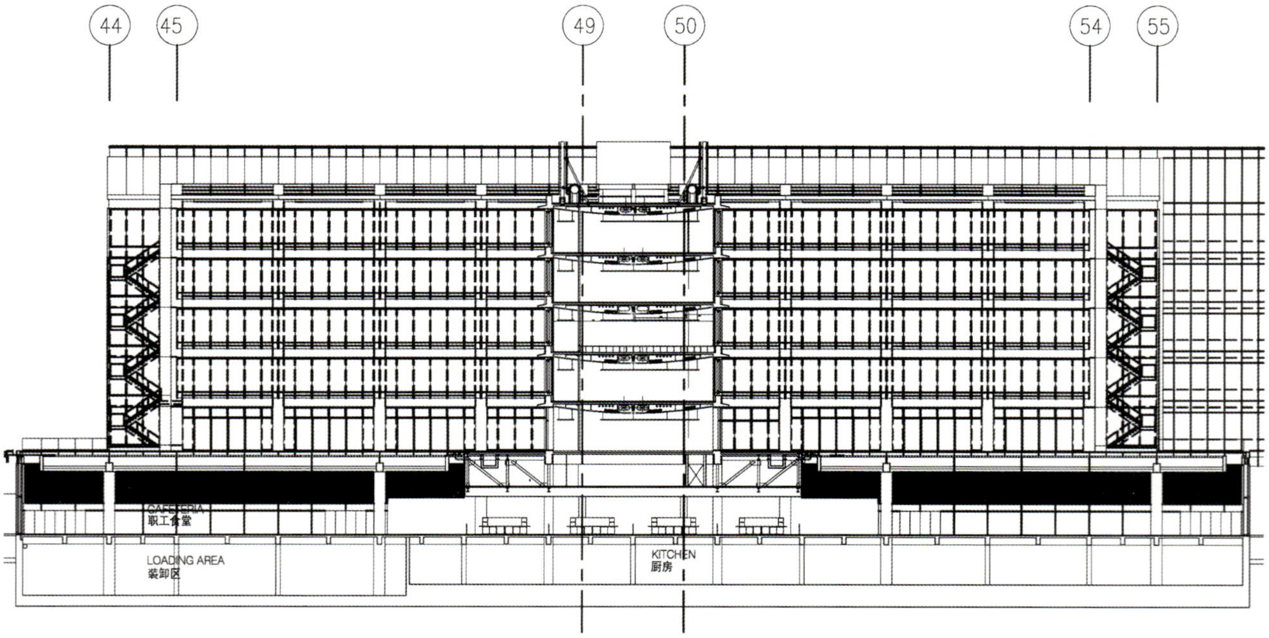
44
45
49
50
54
55
CAFETERIA
职工食堂
LOADING AREA
装卸区
KITCHEN
厨房

中心园区交通便利，无论使用何种交通工具，都很容易到达。地铁线路也将延伸到这里，通过地铁从中心园区将可直达市中心或者更远的地段。

中心园区由总建筑面积超过 30 万平方米的研发空间和一些配套建筑所组成，充分利用了这块面积达 27.5 万平方米的狭长建筑用地。办公大楼楼层不高，设计时充分考虑和利用了朝南向的宽阔地带和所在地的景致。为了减少建筑的狭长感，项目将研发空间设计成了两座线形五层高的侧楼。每个侧楼设有一个中庭，由中央办公区域连接起来，而中央办公区域内设有行政管理办公室、会议室和文娱中心。

园区将容纳 7500 名研发人员从事先进的新一代移动技术的研发。园区建筑内部分为开放式办公室和封闭式实验室。尽管各部门彼此隔离，但是各个区间临走廊一面都有天窗，天窗彼此相连。为了使研发空间在使用时具有最大的灵活性，也为了满足将来研究和技术需要对空间作一定改造的要求，研发楼层被设计成了一系列的 18 米 × 9 米的结构格子间。这种设计使自然光尽可能多地照进格子间内部，同时从里面也可更多地观赏到外面的景致。

沿着宽阔的大楼前廊走进去便是坚固的入口，由此可以乘坐电梯，也可走楼梯到上层的研发办公区域。各研发区域都与过道和走廊相连。位于两侧的用于研发的侧楼在视觉上更加突出了中央的入口门厅。凸出的中央门厅周围是非研发区、行政办公室、产品陈列室、会议室和培训室。中央门厅下面的地下室里是员工餐厅和专门的研究设备储存空间。研发侧楼下面是可容纳 3000 辆车的多层地下停车空间。

在设计研发大楼时，建筑师也充分考虑了这里大范围的自然景观。为了降低较高的地下水位和被水淹没的风险，建筑师将室外设计成了一系列的湿地景观，既有储存雨水的水池，也有用于生物过滤的小溪流。长满当地植物的花园和水景在视觉上与办公大楼互相映衬，也为员工提供了理想的室外聚会和交流场所。

大楼结构体系的设计注重高效能和构件使用的革新两方面。框架结构使用了典型的锥形后张拉预应力框架，这样处理的效能高，使用的材料少，可节约大量的材料。研发办公室模块的建筑和结构设计的可重复性简化了建设过程，从而减少了材料浪费，特别是可重复使用一些模板，使建筑工作本身变得更快、更有效率。压力基桩使用了预应力高强度混凝土，在轴承座的空心管里加入了钢桩靴，节约了混凝土材料。指定使用的可循环材料和水泥替代品起到了保护资源的作用。

整个中心园区被规划成当代高效能的办公环境，其外部幕墙采用了透风、双层内嵌绝缘不反光低辐射玻璃系统和铝板材料。在双层玻璃间还嵌有机动百叶片，百叶片与自动传感控制装置连接，以调节和平衡计算机为主导的工作环境下的光度。该项目使用的玻璃都经过精挑细选以使建筑内的景观、热舒适度和亮度之间达到平衡。窗户可手动操作，方便进行自然通风、排烟和清洗。因为玻璃腔内形成了绝热层，双层幕墙的设计也降低了建筑本身的热负荷。

18 米宽的拱廊设计既为建筑提供了充足的自然光，也降低了沿途研发功能体的照明负载，同时还提供了聚会场所和横穿整个项目的过道。拱廊内的温度受办公室排出的热的影响而变热或者变冷。水景式花园和竹林相互辉映，为建筑群的脊梁——拱廊提供了天然的背景。总而言之，封闭的机械、电气和管道提供了横向稳定性，外墙玻璃里内嵌的遮阳设施以及预制混凝土屋顶等元素将上海华为研发中心园区的多个建筑统一成具有一致身份和持久特性的综合建筑群。

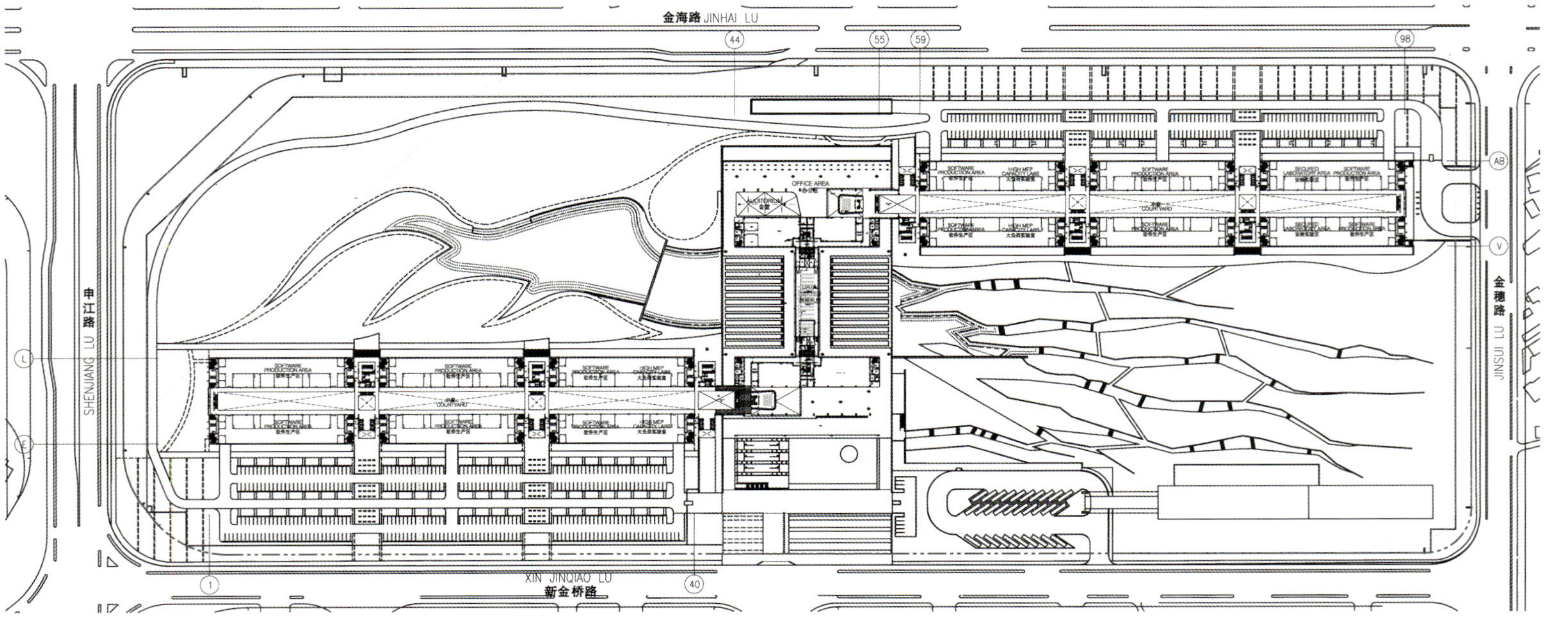

金海路 JINHAI LU
44
55
59
98
OFFICE AREA
AUDITORIUM
SOFTWARE PRODUCTION AREA
HIGH MEP CAPACITY LABS
SECURED LABORATORY AREA
COURTYARD
申江路 SHENJIANG LU
金穗路 JINSUI LU
XIN JINQIAO LU
新金桥路
40

TT Project

About the historical and urban context

The architect was approached to construct a new building at the site of the old Hankook Ilbo building originally designed by the renowned Korean architect Kim Swoo-geun in 1968. The area had recently been demolished as part of an urban renewal development project. The site is well-known to the public due to the close proximity to Gyeongbok Palace and Insa-dong, both landmarks of Seoul. Given this historical context, the architect personally believes it is inappropriate to destroy older sophisticated buildings built in the 1960s and 1970s and replace them. However, given the inevitable situation involving the government's urban redevelopment project, the architect had no choice but to accept the situation, then decided to design in consideration of the historical and urban context of the site.

The specific architectural shape was made in response to the distinctive geographical features. The site consists of several acute angles, and within its figure, inducing interactions between horizontal and vertical dynamics, such as water flowing in various directions. This recalls a birch tree trunk that weathered severe hardships over time.

Directly south of the building an area with historical commercial alleyways, one famous called Pimatgol. Although the area has largely disappeared with the urban renewal project, it is still possible to connect with some of the remaining traces. The connection made through the site is directed toward Dongsipjagak, an ancient watch tower at the southeast corner of Gyeongbok Palace. This area connects to Samcheongdong, an area which remains vibrant, evoking both the historical tradition and urban context.

ARCHITECT
Cho Byoung-soo (Principle)

FIRM
BCHO Architects Associates

PROJECT TEAM
Kang Woo-hyun, Erik Horn, Park Gi-hyun, Pai Yong-eun,Nicholas Locke, Kwon Sung-Hwa, Seo Ji-young, Park Joo- hyun, Eric Druse

STRUCTURE
Post-Tension system, SRC

CONTRACTOR
Hanil Engineering & Construction Co., Ltd.

STRUCTURAL ENGINEERING
Dongyang Structure Engineering

FAÇADE CONTRACTOR
Lucky Ability - CurtainWall - Seoul, Korea

FAÇADE CONSULTING
Simpson Gumpertz & Heger - New York City, USA

SERVICE ENGINEERING
Han Kook Engineering Co.

AREA
54918.85 m^2

LOCATION \ 14, Junghak-dong, Jongno-gu, Seoul, Korea

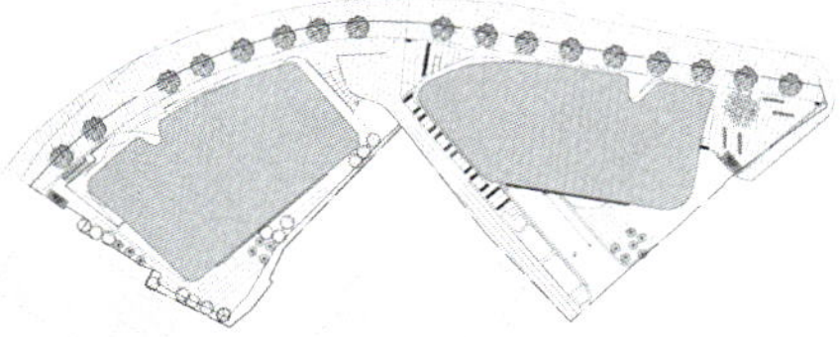

해우리

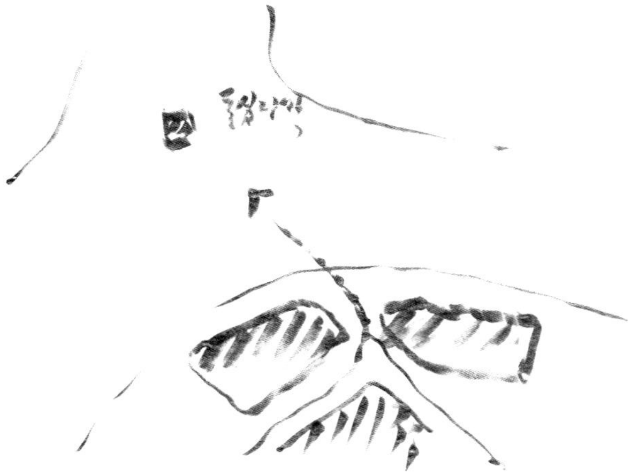

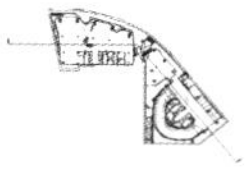

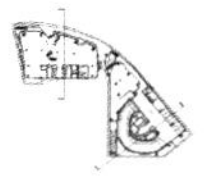

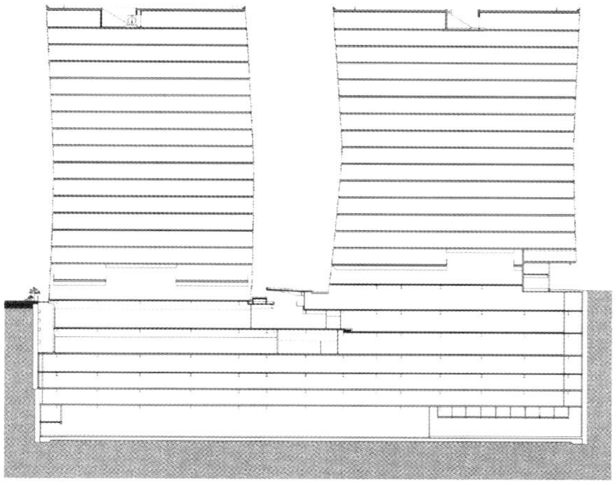

A-A' 단면도

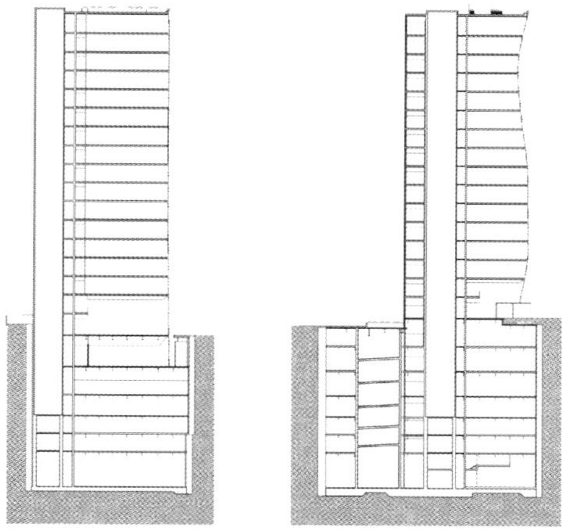

B-B' 단면도

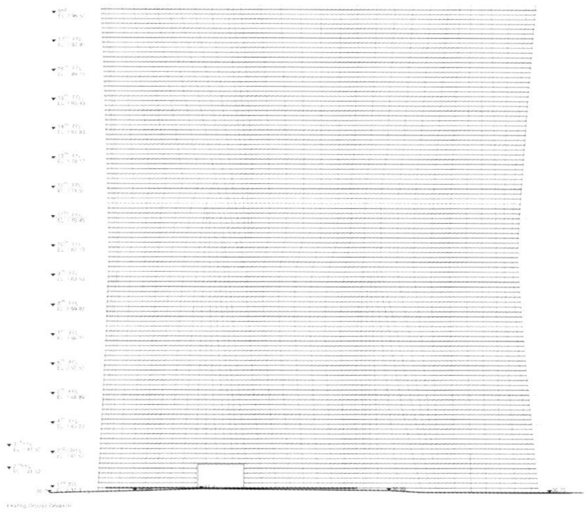

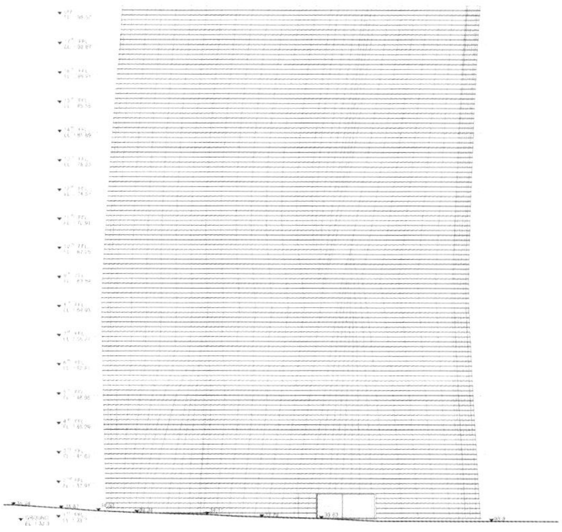

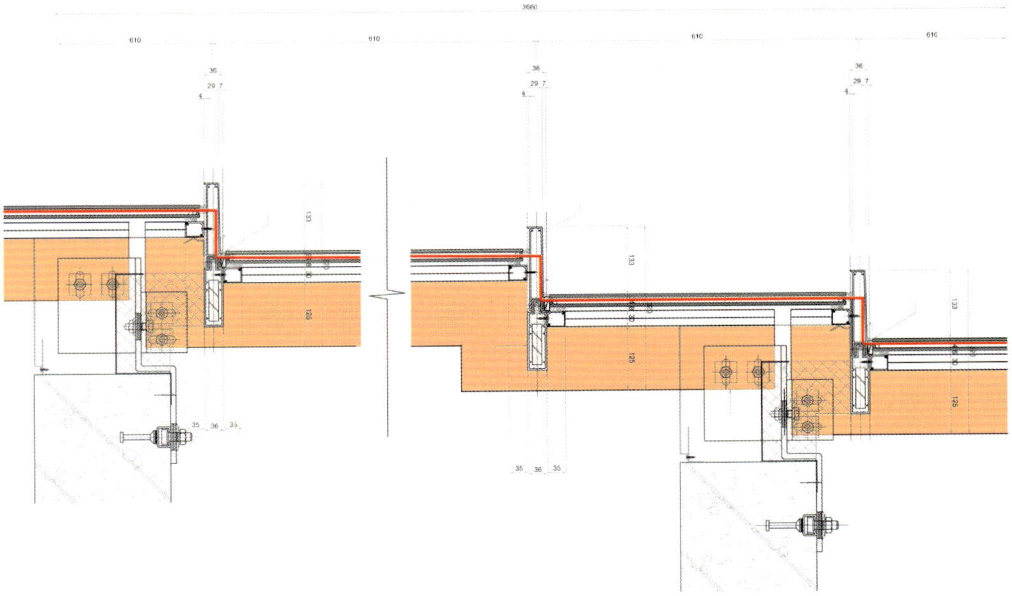

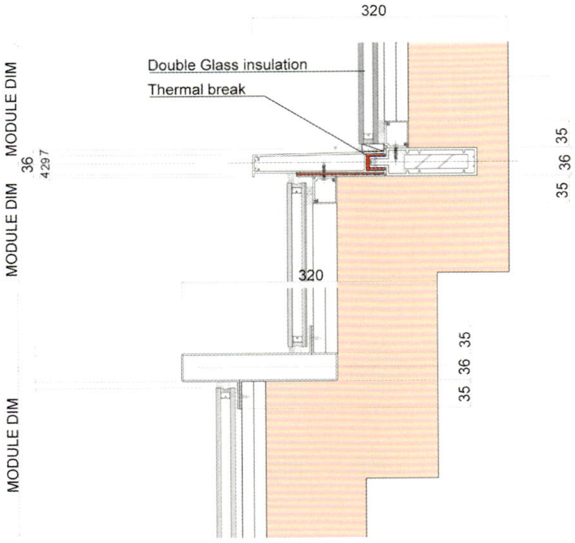
320
Double Glass insulation
Thermal break
MODULE DIM
36
4297
MODULE DIM
320
35
36
35
35
36
35
MODULE DIM

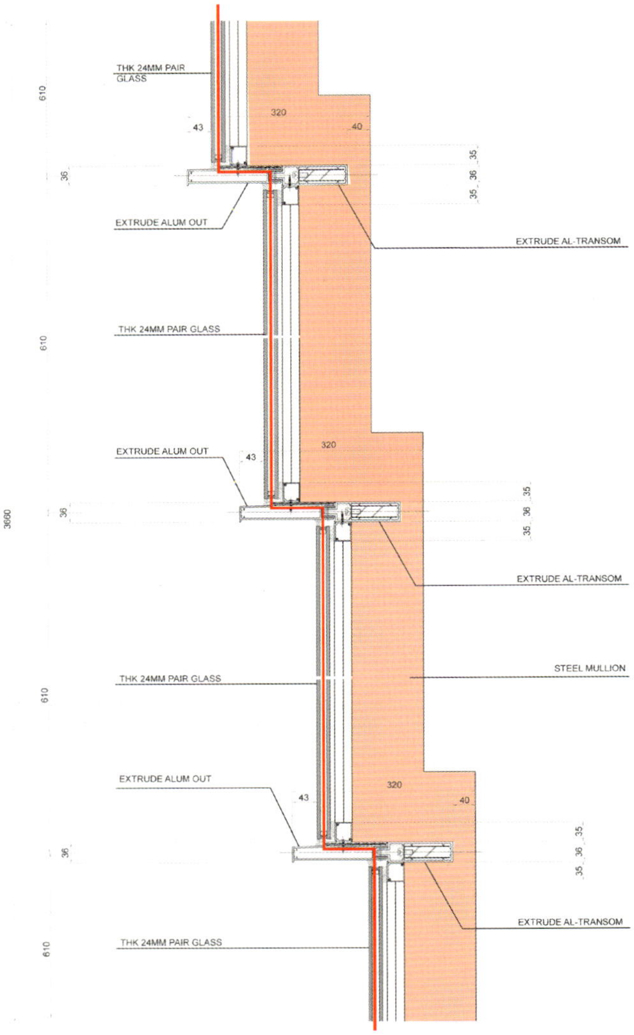
THK 24MM PAIR GLASS
EXTRUDE ALUM OUT
EXTRUDE AL-TRANSOM
THK 24MM PAIR GLASS
EXTRUDE ALUM OUT
EXTRUDE AL-TRANSOM
STEEL MULLION
THK 24MM PAIR GLASS
EXTRUDE ALUM OUT
EXTRUDE AL-TRANSOM
THK 24MM PAIR GLASS

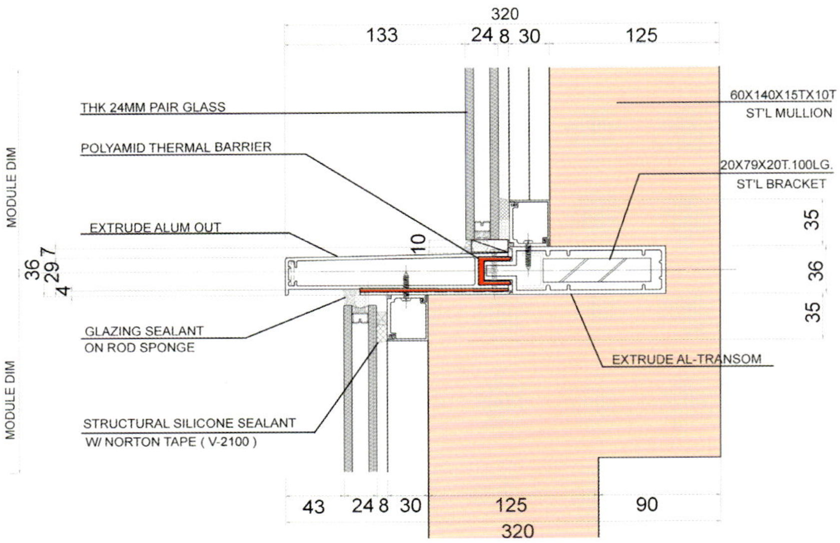
320
133
24 8 30
125
THK 24MM PAIR GLASS
60X140X15TX10T
ST'L MULLION
POLYAMID THERMAL BARRIER
20X79X20T.100LG.
ST'L BRACKET
EXTRUDE ALUM OUT
MODULE DIM
36
29 7
4
10
35
36
35
GLAZING SEALANT
ON ROD SPONGE
EXTRUDE AL-TRANSOM
MODULE DIM
STRUCTURAL SILICONE SEALANT
W/ NORTON TAPE (V-2100)
43
24 8 30
125
90
320

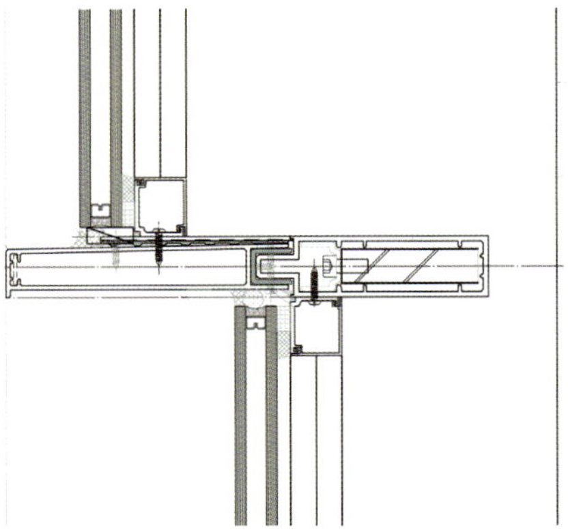

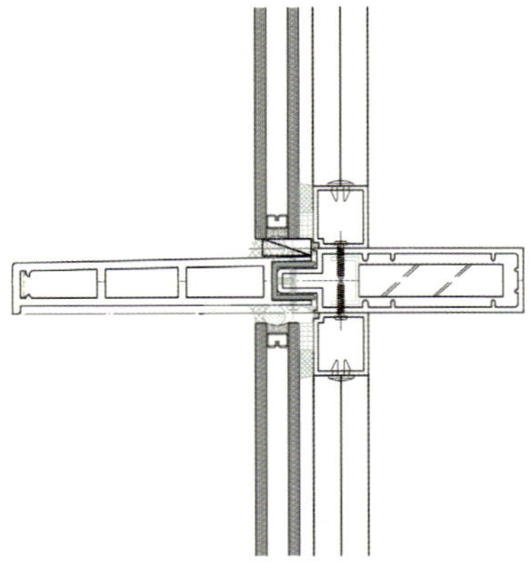

RMATION 안내
합뉴스

A nurbs* surface modeled to evoke the strength and primordial nature of the twisted tree.

Development of NURBS (Non Uniform Rational Basis, or Bézier Spline) began in the 1950s by engineers who were in need of a mathematically precise representation of freeform surfaces like those used for car bodies and ship hulls, which could be exactly reproduced whenever technically needed. Prior representations of this kind of surface only existed as a single physical model created by a designer.

from Wikipedia

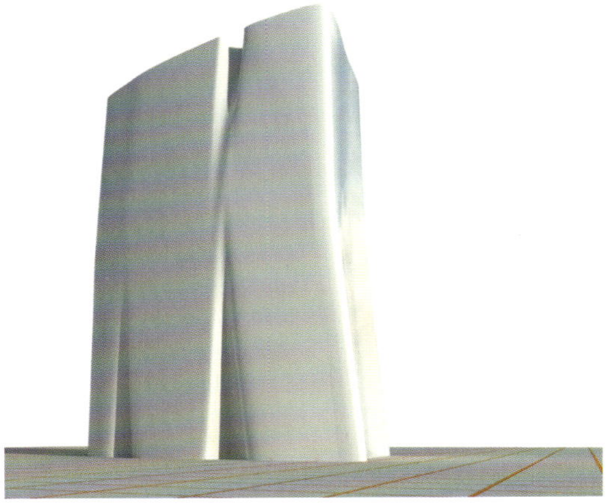

Once again the surfaces are trimmed.

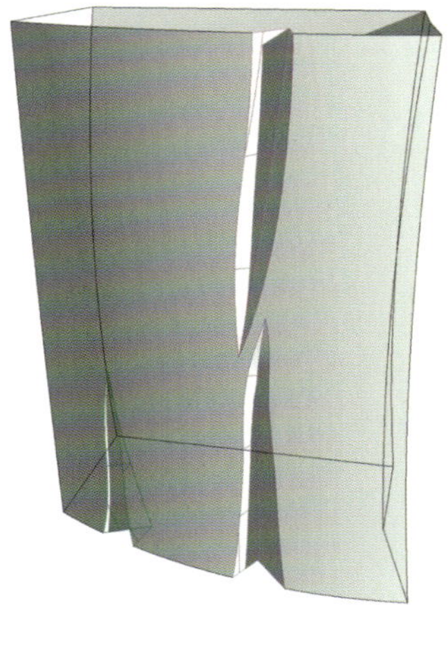

The construction lines for the "notches" are lofted.

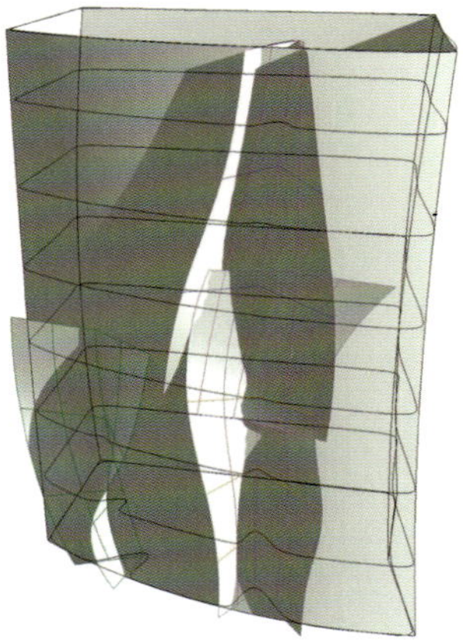

The filleted sections are extruded to reveal shape of the 600mm tall panels of glass that will make up the facade. At this point the length of glass as well as the radius required for construction can be calcuated.

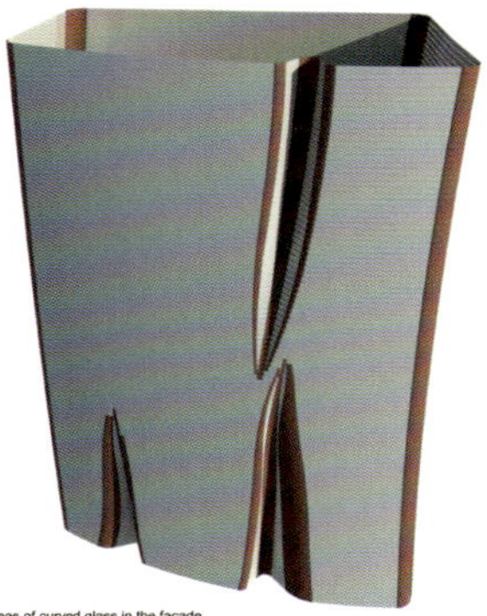

The red areas illustrate areas of curved glass in the facade

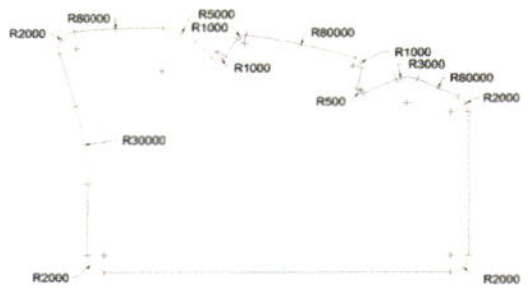

Measurements of a section generated from the "simplified" surface verify that the radii have remained constant from the bottom of the building to the top

Free form surface must be refined into into pure arcs and lines. To achieve this we first create section lines of the n.u.r.b.s. surface.

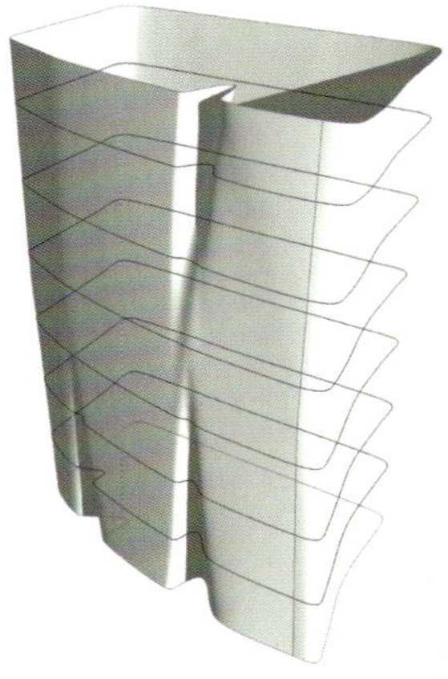

By analyzing these sample sections we can see the basic form of the building can be defined by the top, bottom, and the middle sections.

The top, bottom, and the middle sections are traced with circles and lines. This creates a geometry that can be constructed with standardized modules.

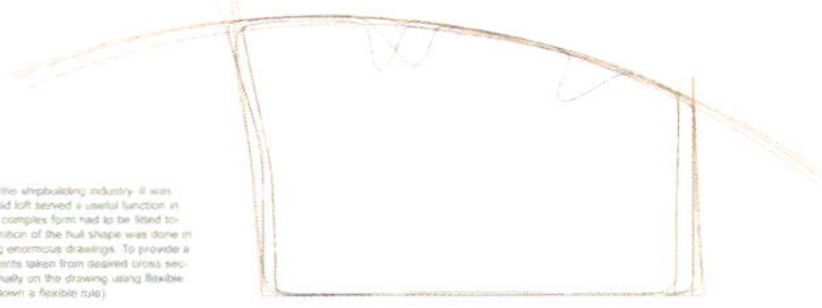

"The name lofting comes from the shipbuilding industry. It was found that work done in the mold loft served a useful function in constructions where parts with complex form had to be fitted together to build a ship. The definition of the hull shape was done in the loft over the shipyard, using enormous drawings. To provide a smooth longitudinal contour, points taken from desired cross sections were connected longitudinally on the drawing using flexible lead "ducks" (weights holding down a flexible rule).

Source:
LOFTING AND CONICS IN THE DESIGN OF AIRCRAFT

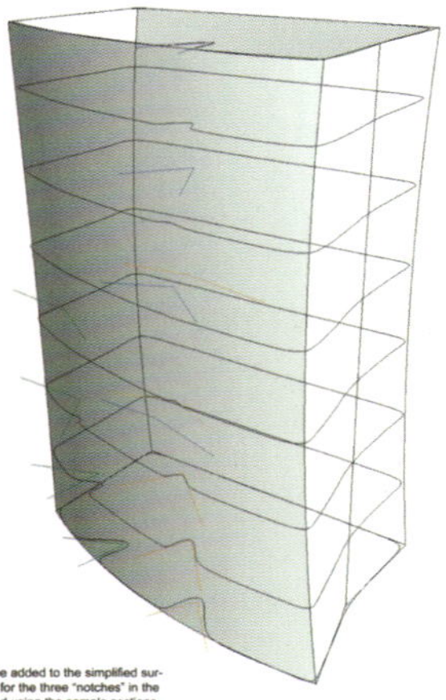

At this point more detail can be added to the simplified surface model. Contruction lines for the three "notches" in the face of the building are created using the sample sections as guides

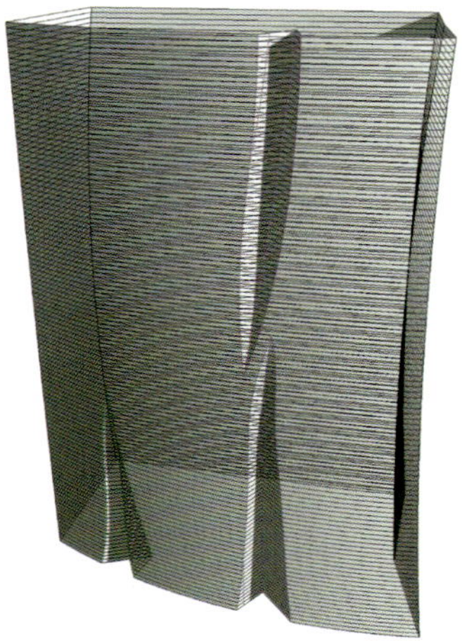

Now that the "simplified" surface has sufficient detail sections can be created from it that will inform the final construction. We have created a horizontal section every 600mm starting at the base of the surface and continuing to the top. This will be the Height of each level of glass that make the facade of the building.

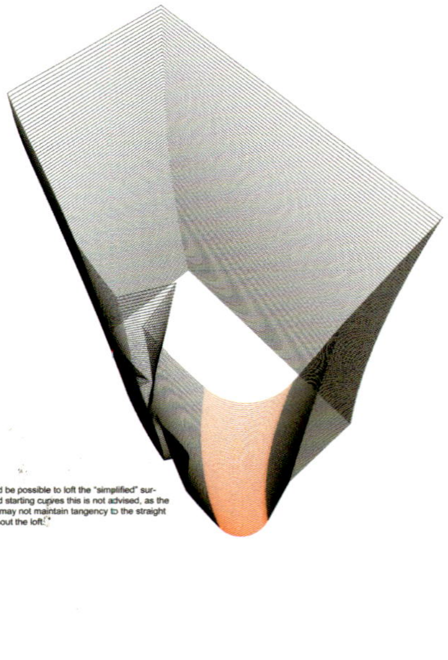

Although it would be possible to loft the "simplified" surface from filleted starting cupres this is not advised, as the curved sections may not maintain tangency to the straight sections throughout the loft.*

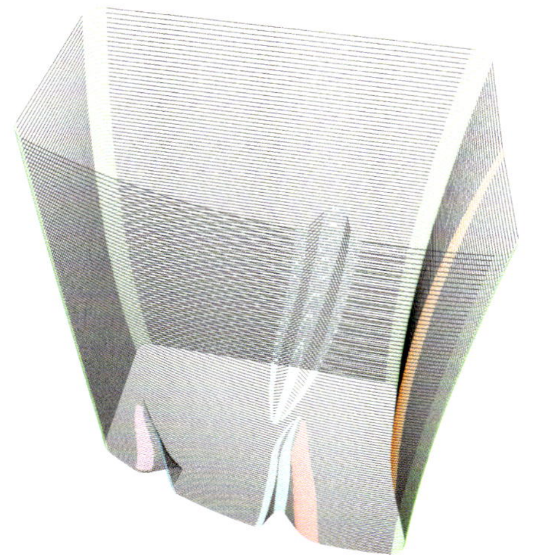

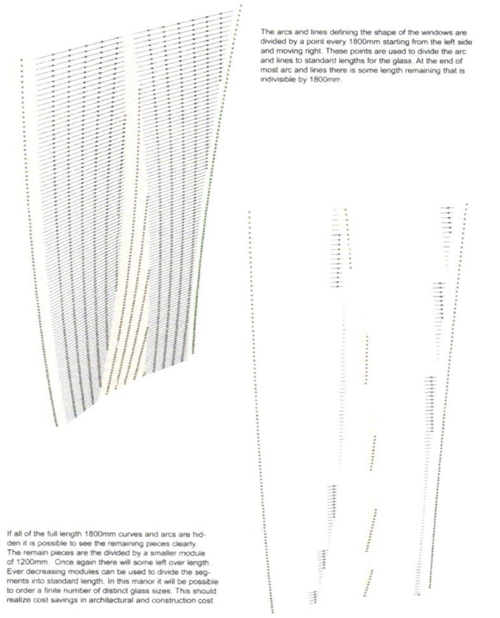

The arcs and lines defining the shape of the windows are divided by a point every 1800mm starting from the left side and moving right. These points are used to divide the arc and lines to standard lengths for the glass. At the end of most arc and lines there is some length remaining that is indivisible by 1800mm.

If all of the full length 1800mm curves and arcs are hidden it is possible to see the remaining pieces clearly. The remain pieces are the divided by a smaller module of 1200mm. Once again there will some left over length. Ever decreasing modules can be used to divide the segments into standard length. In this manor it will be possible to order a finite number of distinct glass sizes. This should realize cost savings in architectural and construction cost.

Here is a proposal for modular glass sizes for the facade. This wall shows four sizes of glass each for four radii of glass for a total of 16 glass modules. Evident in this image are leftover gaps which require further resolution.

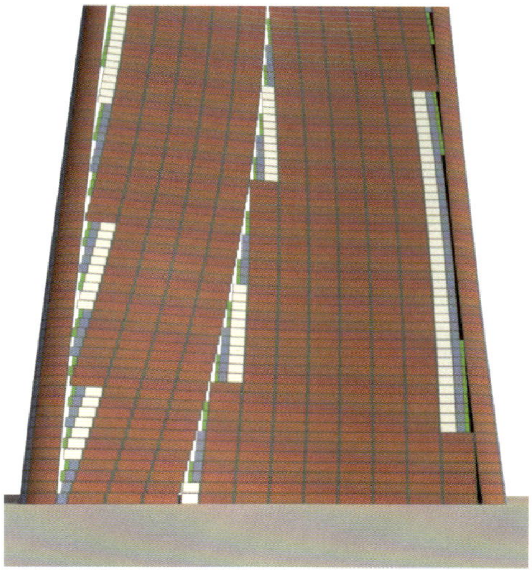

历史背景和周围城市环境

金寿根于 1968 年设计建造的原韩国日报大厦在城区改造中被拆除。建筑师几乎相当于在原日报大厦地基上重新建造了这个工程。因其与韩国著名的历史建筑，也是首尔的地标工程景福宫和仁寺洞相邻而闻名。鉴于其特殊的历史背景，建筑师个人认为将建造于 20 世纪六七十年代的古老而精致的建筑全部拆除而取而代之不太合适。然而政府实施的城区改造工程却让旧有建筑被拆除无可避免，他别无选择，只能接受。在设计的时候建筑师决定考虑该址的历史背景和周围城市环境。

地基由许多尖锐不规则的角区组成，建筑形态的设计充分考虑了地基独特的地理特色。其在垂直和水平方向的动感表达互相呼应，宛如流动的泉水充满灵性，也让人想起饱经岁月风霜的古老的桦树干。

紧邻该项目南面的是历史上的商业小巷，其中一条著名的小巷叫 Pimatgol。由于城区改造，很多古迹业已消失，但是该建筑还是与市区残余古迹相连。穿过大楼就可到达景福宫西南角的古瞭望塔 Dongsipjagak。另外，该建筑还与三青洞相邻。灵动的建筑使此独特的地理位置更加充满活力，游客在此既能感受到现代都市的气氛，也能瞻仰历史古迹。

Here are three different colored sets of arcs. Each color represents a fixed radius. By filleting the disinct straight sections in this maner it is possible to limit the number of different curvatures required for the facade glass from infinate to six or seven.
Notice that when ratio of the angle of the intersection of two lines to the radius of the circle becomes too small it is imposible for a fillet to form. This leaves a discrepency in the surface the must addressed in detailing or by altering the geometry of the "simplified" surfaces.

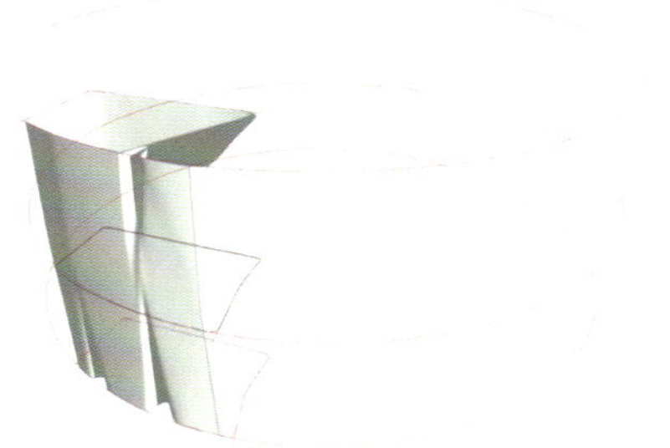

The standardized curves approximate the original form

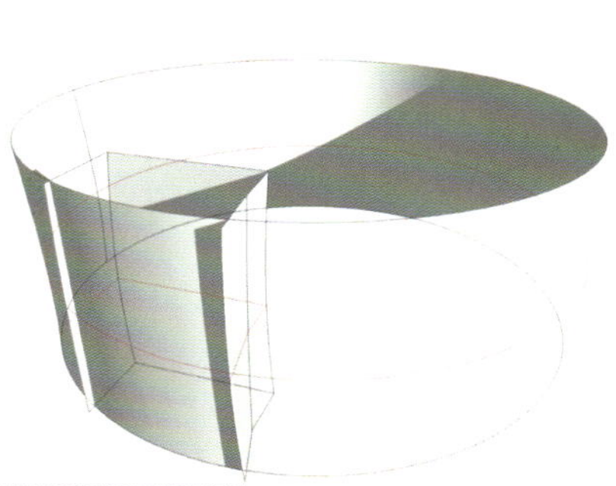

By lofting" the standardized curves a surface is created that approximates the original

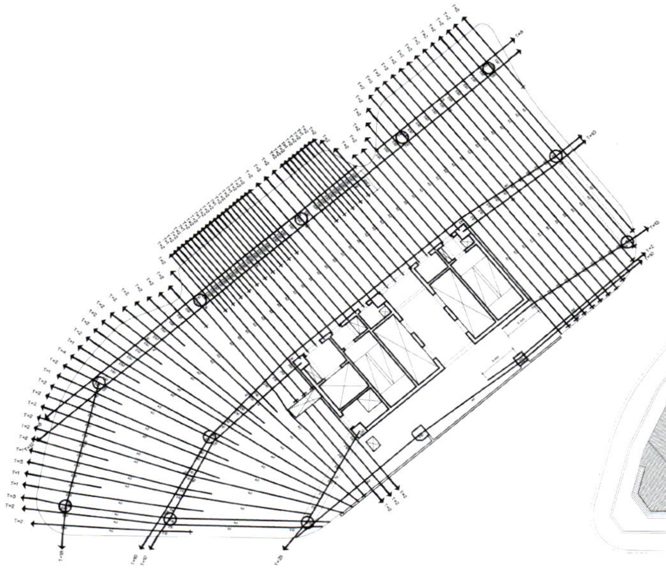

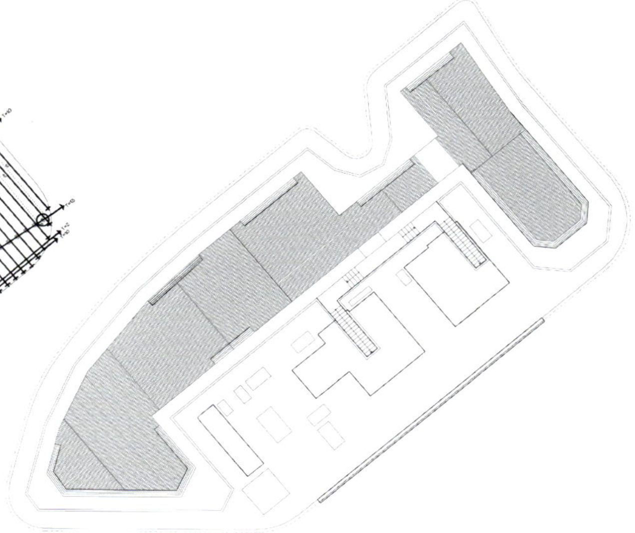

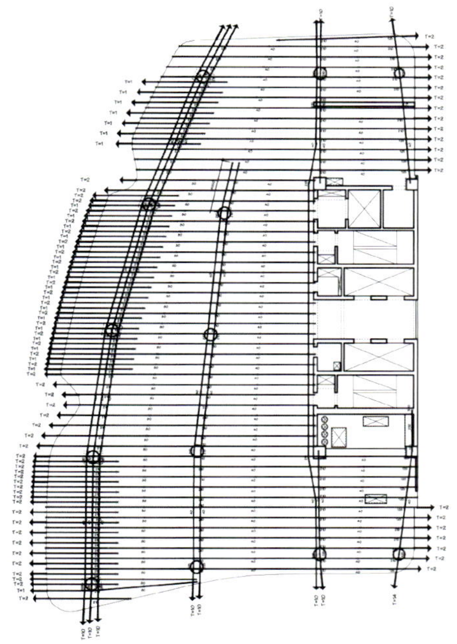

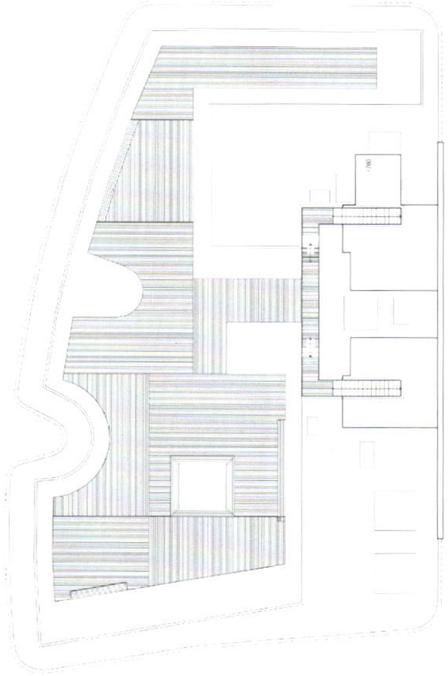

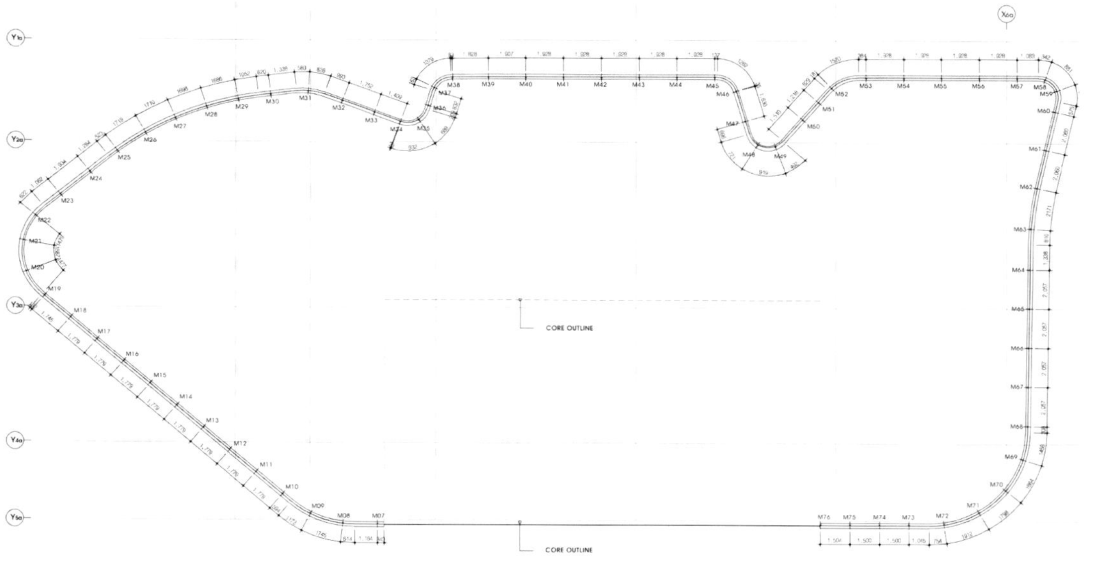
CORE OUTLINE
CORE OUTLINE

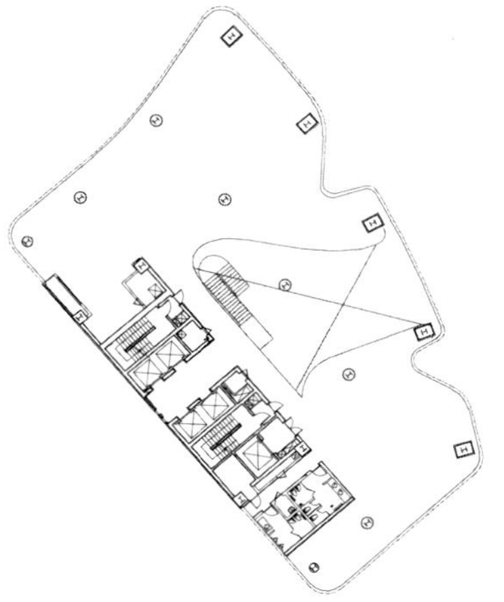

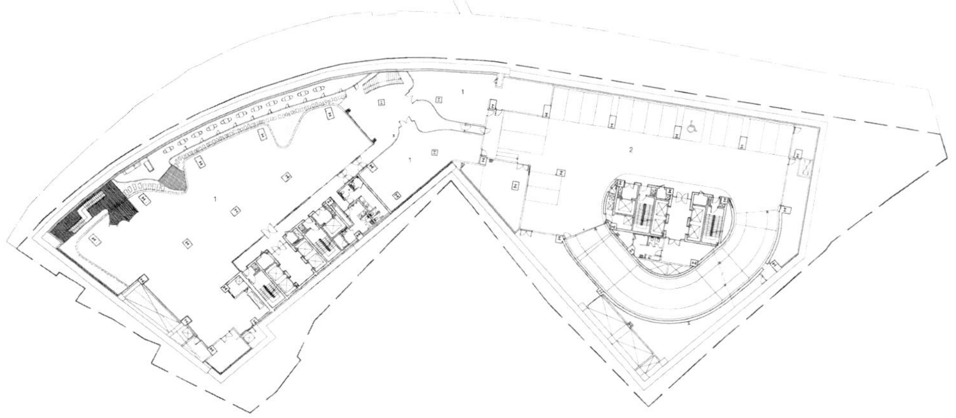

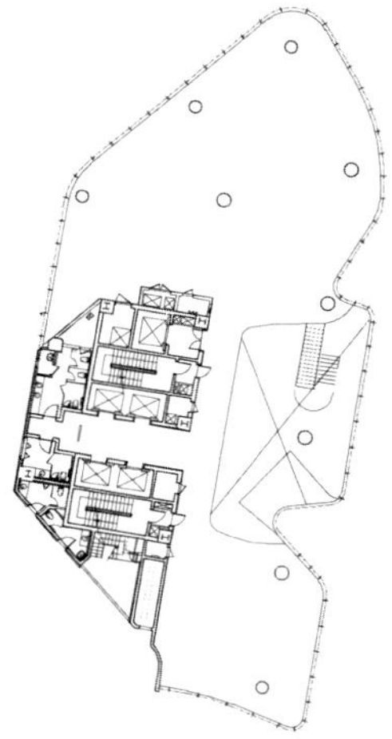

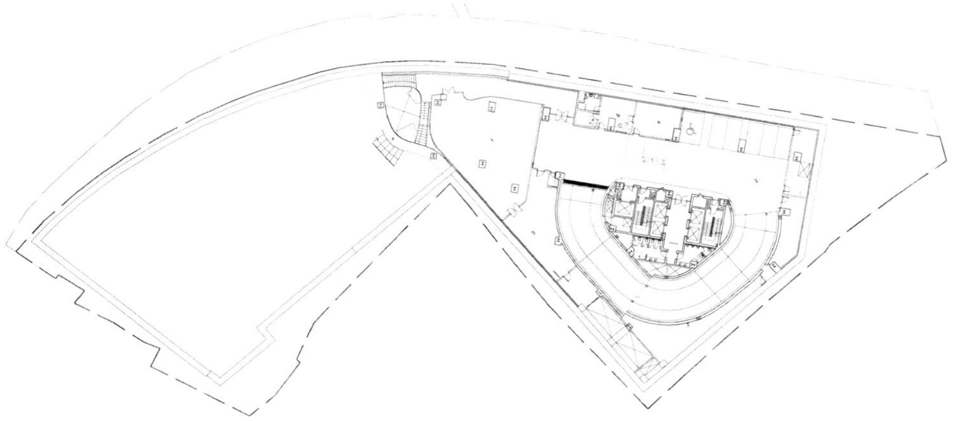

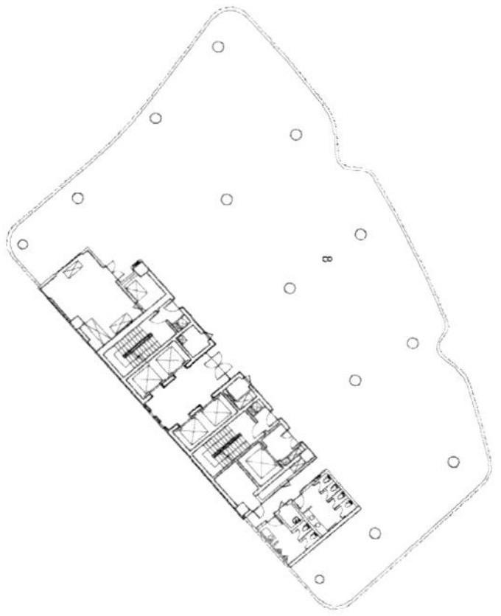

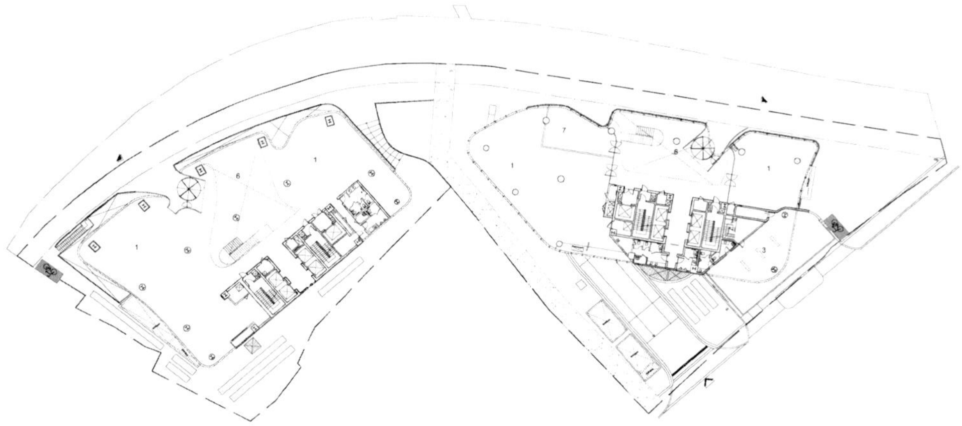

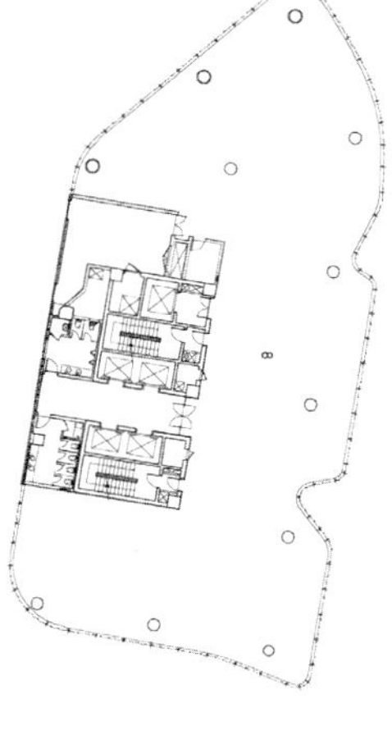

Sequana Tower / Bouygues Telecom Headquarters

The project is organized around the power of a tower located at the centre of the composition. It rises in the same location as the existing EDF tower, yet expanding its footprint and height to maximize the volume within permitted envelope.

The elliptical form gives the form direction and multiple fronts. There is no front or back, yet its directional quality conveys movement from Issy to Paris, from the streets to the peripherique, from land to river. Its soft shaped edges avoid hard edges and corners, implying a democratic space and certain sensuality in its relationship to the green landscape that surrounds the building at its base.

The elliptical prism is eroded as it reaches the sky, following the angles of the envelope. This erosion results in two of angled planes that reveal the curved plan in elevation. The receding floor plates provide additional space that can be captured for functional use. They also introduce executive floors with smaller footprints and angled glass surfaces that act as greenhouses and add uniqueness to the top floor spaces.

The tower is organized around a center core of elevators, stairs, technical rooms and toilets. The balance of the center space is dedicated to meeting rooms and office support functions. The elliptical form allows maximizing the amount of office space by widening in the centre to accommodate the core and narrowing towards the ends to minimize interior space. The result is a large floor plate area with maximized perimeter offices within the permitted 75 meters maximum length.

The site also contains a low-rise building that accommodates the balance of the permitted buildable area. A bar is placed along the street running parallel to the ellipse. It bends to form a sinuous S curve that responds to the force of the ellipse. Its plasticity adapts to the sensuality of the ellipse. The S bar contains more classic offices either at 18 meters in width or two zones of 12 meters along a linear glass spine.

A third building exists across the Paris boundary. This building will be restored and adapted for modern office use. The three buildings are connected by an atrium that rises from the landscape like a glass wave. This form will act as the arrival lobby between the tower and the S bar. Its form translates the plan aspects of the tower and bar into a sectional expression that engages the surrounding landscape.

All three building forms are inspired from nature. Their organic message is projected subtly and mysteriously. They have organic overtones yet they are clearly sculpted and shaped by the technological man. The tower is inspired on a river rock, the bar on a twig, the atrium on an earth form. They are symbols of the green society. Yet in the quest for an expression of the technological era, the forms are abstracted to their essence and wrapped in the modernity of glass and steel. At the urban design level, the composition addresses the duality of the site. The building form faces both Paris and Issy. So do the entrances to the atrium. And the bar acts as a physical and visual link from the river to the city. The open space flows through the site and past the atrium to link visually the perpherique and the street. The approach by car from the quay is equalized with the approach from the public transportation.

ARCHITECT
Bernardo Fort Brescia, Laurinda Spear

FIRM
Arquitectonica

AREA
45000 m^2

PHOTOGRAPHER
Paul Maurer

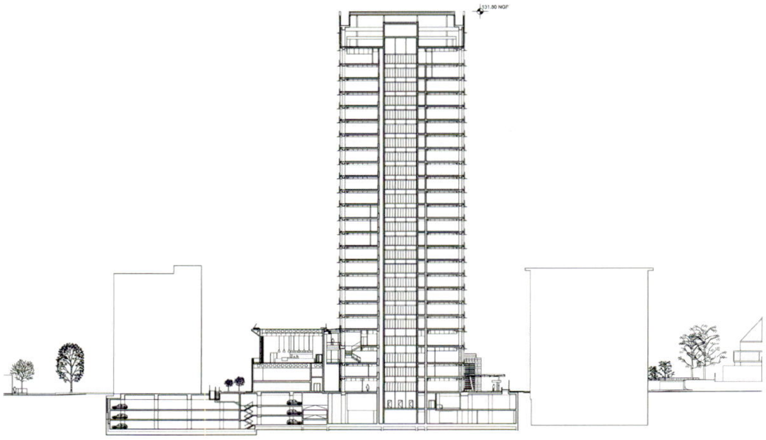

SECTION
0 5 10 15 20 25M

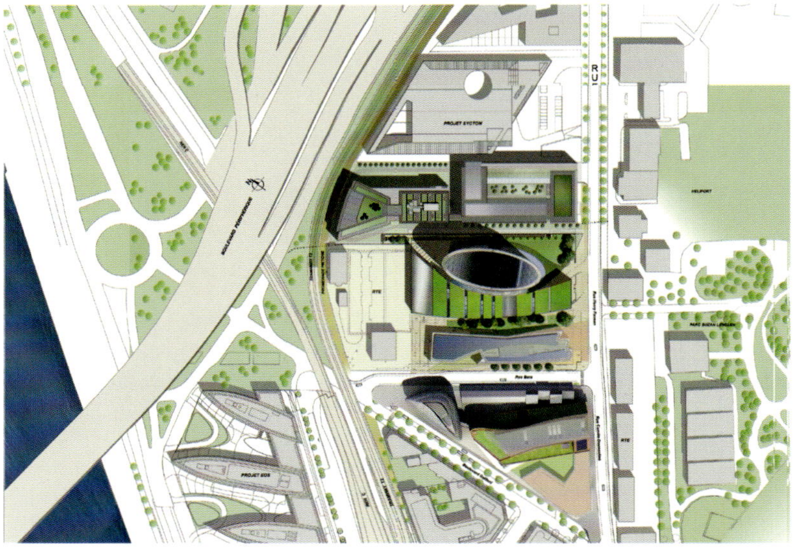

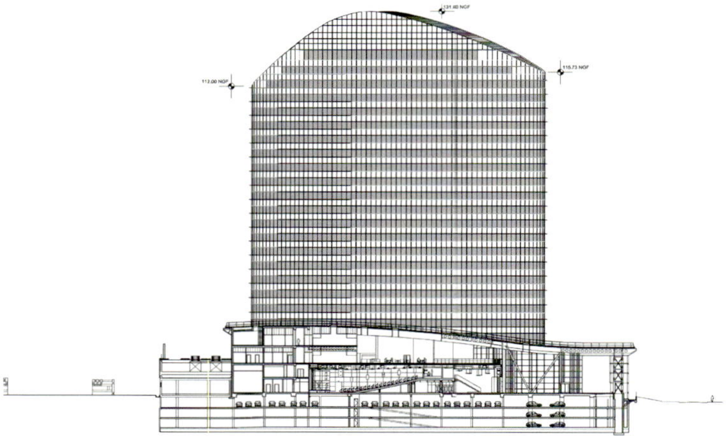

SECTION
0 5 10 15 20 25M

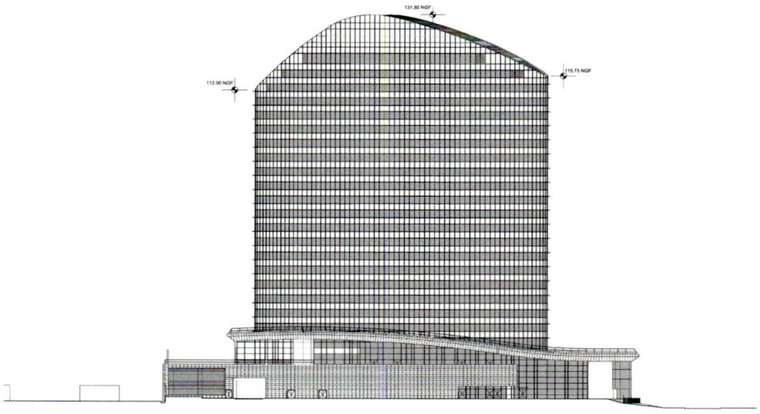

WEST ELEVATION
0 5 10 15 20 25M

bouygues

该项目是围绕位于组合体中心的塔楼进行的。它与法国电力公司塔楼处于同一地点，但却最大限度地进行了横向和纵向的扩展。

椭圆形既指出了方位，也对建筑物进行了多方面的正面描绘。整个建筑物外观上没有前后之分，而其方位特性传达的是一种运动感，从伊西到巴黎，从街道到四周，从陆地到河流，无处不在。其柔和的边界造型避免了硬边和角落的产生，具有一定的观赏性，并与建筑物底部的绿色景观融为一体。

椭圆棱柱形建筑物在其顶部沿其外围方向被削掉了一部分，形成了两个直角的平面，立视图则表现为两个曲面。渐退式的楼板提供了更多可使用空间。同时也提供了小面积的行政办公楼层，直角玻璃面可充当温室，使顶层空间别具一格。

这座塔楼围绕一个中心轴，有电梯、楼梯、技术室和厕所。中央空间设有会议室和办公区。椭圆形设计通过拓宽中心面积，缩小边缘以减少室内空间来确保办公空间的最大化。于是长达 75 米的最大周长办公区域楼面由此产生。

这里还包括一个低层建筑，使可使用的建造区域达到平衡。沿街有一个隔间，与椭圆形建筑平行。隔间弯曲成 S 形，与椭圆形建筑交相辉映。隔间的弯曲性与椭圆形建筑的观赏性互相协调。S 形隔间中有着更多的典型办公空间，有一个宽 18 米的办公空间，还有 2 个长 12 米的办公室，都是沿着这个椭圆形玻璃建筑一字排开的。

第三座大楼位于巴黎的边界。这座大楼将进行修复以作为现代办公室之用。这三栋建筑由一个中庭连接，中庭犹如玻璃形成的海浪。中庭将作为到达塔楼和 S 形隔间的走廊。这种形式既诠释了在设计中塔楼和隔间之间的组合关系，又使它们与周围的环境相吻合。

三个建筑作品的设计灵感均来源于自然界。它们蕴含的几何信息被巧妙地传达出来。技术人员的精心刻画与设计也掩盖不了建筑物的几何色彩。塔楼的灵感来自于河边的岩石，隔间的灵感来自于树枝，而中庭的灵感来自于地球的形状。这些都是绿色社会的象征。为了表达科技时代，设计形式被抽象化，由现代化的玻璃和钢材所包覆。在城市规划方面，作品强调了选址的二重性。建筑物一面对着巴黎，一面对着伊西。中庭的入口设计也是如此。隔间充当着一个有形的、可视的连接体，把河流与城市连接起来。开放空间穿过选址地并经过中庭与周围景观和街道相连。从码头开车和乘坐公共汽车到这里所花的时间差不多。

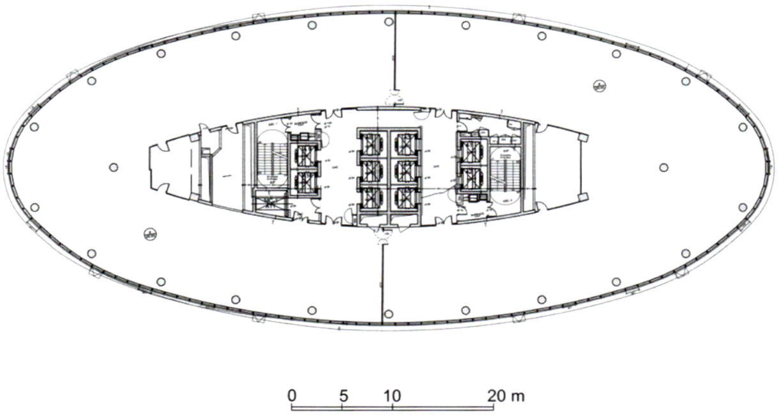

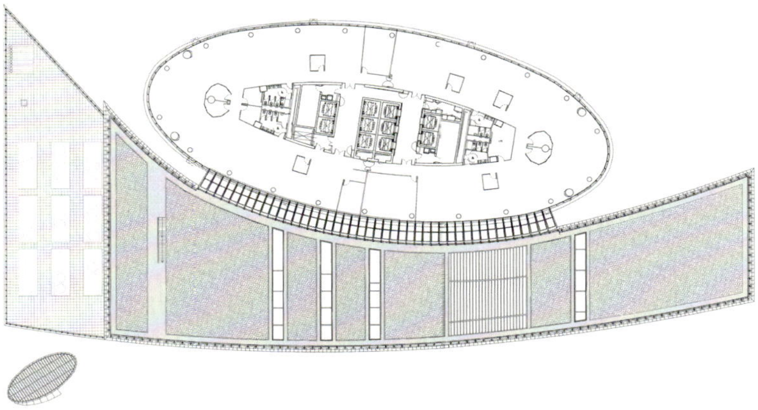

TYPICAL FLOOR
01 3 6 10M 15

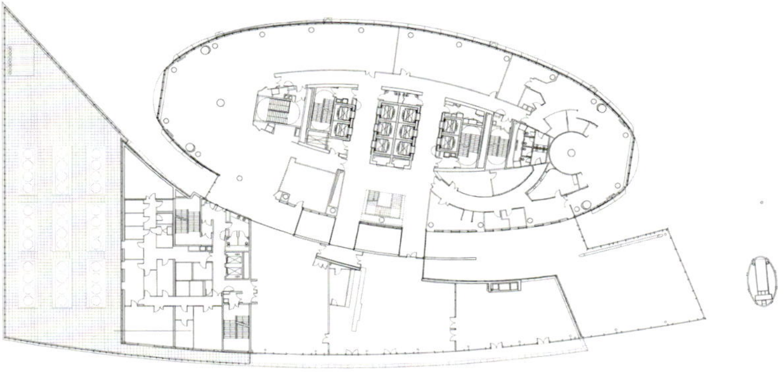

LOWER FLOOR
01 3 6 10M 15

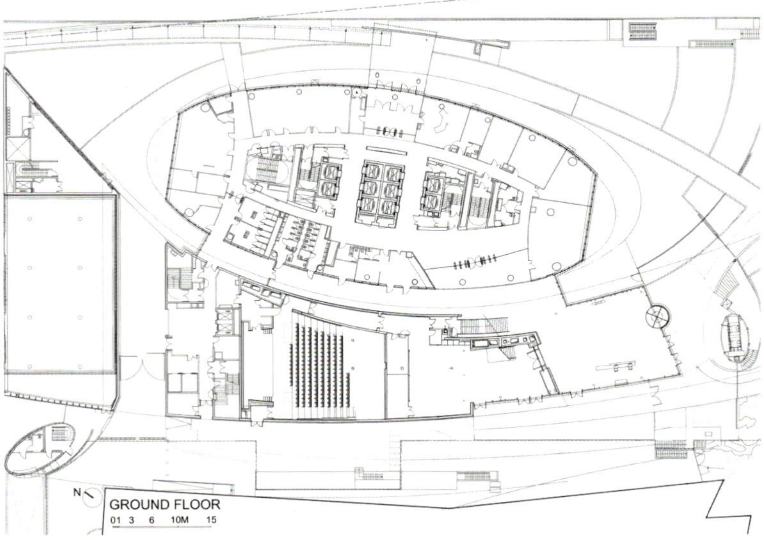

GROUND FLOOR
01 3 6 10M 15

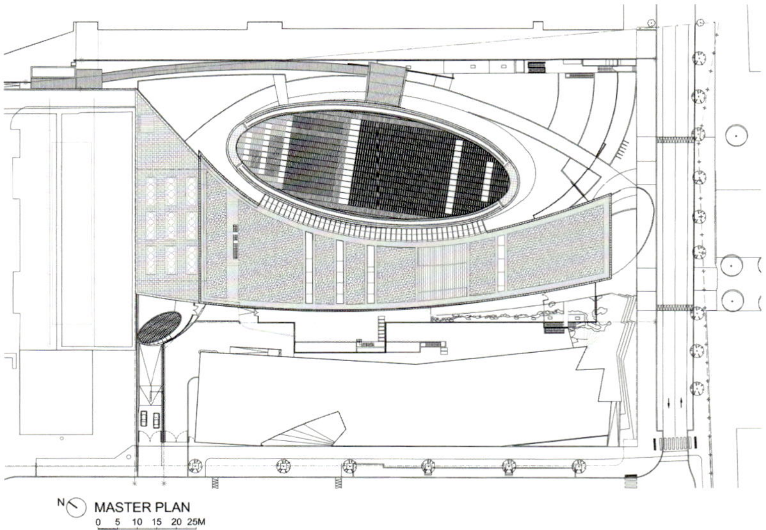

MASTER PLAN
0 5 10 15 20 25M

INTERFACE Office Building

For half a century, the site of the INTERFACE building was occupied by the Netol cleaning products factory. The building designed in 1955 by the architect Santiago Balcells was an industrial complex of warehouses built against each other. The façade gave onto two streets, Tànger and Ciutat de Granada, and comprised a base of stone cladding with vertical concrete slats above The main entrance was framed in stone, forming an independent element on a larger scale.

Urban planning for the street block stipulated that the site was to conserve the old factory façade and allowed building levels of 18,271m^2 above grade. In terms of built volume, it proposed the construction of a tower block with a clearly rectangular floor plan aligned along Carrer Ciutat de Granada and a base similar in height to the one in the existing façade.

Our project included the proposal to conserve the façade but eliminated the base, leaving it suspended from the new building's structure. This enabled us to relate the ground floor premises directly with the street and design a new main entrance to the building via an inner courtyard located at the corner. This courtyard leads to the main foyer, a double-height space that communicates with the vertical communication shafts.

The proposed built volume applies the criteria of the preliminary planning study that sets out to relate the building with others on the same block and in neighbouring sectors, such as the Audiovisual Campus. The idea was to make part of the building the same height as the old façade, thereby establishing a relationship with the adjacent building. This formed the base of a prismatic 13-floor volume aligned with Carrer Ciutat de Granada. The tower distorts its floor plan to adapt to the corner and accentuate its slender line.

To tone down the volumetric disparity of the buildings, we worked with a continuous skin that envelops the volumes indiscriminately, folding to form a second, scale-like skin at points where extra solar protection is needed. The second skin also envelops the old façade, incorporating it into the complex and making it an inseparable part of the new building. To highlight the new entrance, this double skin is stencilled with the building's name.

ARCHITECT
Enric Batlle, Joan Roig

FIRM
Batlle i Roig arquitectes

COLLABORATOR
Goretti Guillén, Helena Salvadó, architects, G3, technical architecture, STATIC, Gerardo Rodríguez, engineer, structure, PGI Grup SL, installations

AREA
27965 m^2

PHOTOGRAPHER
Eva Serrats

LOCATION\22@, Barcelona, Spain

INTERFACE办公大楼所在的地址原是Notel洗涤用品厂，该厂在该址有半个世纪的历史。由建筑师Santiago Balcells于1955年所设计，洗涤用品厂的建筑是一个彼此连接的仓库工业园区。这些建筑的立面呈现在相邻的两条街道上——Tànger街和Ciutat de Granada街。立面由石头建成，石墙外面覆盖有垂直的混凝土板。其大门是石制的，在一大片建筑群里形成独立的元素。

改建时城镇规划规定，在这里改建建筑必须保留旧工厂建筑立面的原貌，允许在原立面之上建造建筑面积达18271平方米的楼层。在空间布局方面，规划提议建一座塔楼并且平面布局要沿着Ciutat de Granada街成明显的矩形，房基的高度可与原来立面的高度大致相同。

规划根据城镇规划保留原有立面，但是淘汰了基墙，让留下的立面悬在新建的楼体上。这保证了底层直接与外面街道的设置布局风格互相呼应，协调一致；这种设计也使建筑师可以协调地设计一个穿过拐角处内庭的大门。内庭通向两层高的玄关，而玄关通向垂直升降机井。

最后建成的建筑符合设计规划时的标准，与同一条街上的建筑风格相容，并与临近的视听大学校园相呼应。新建建筑地基部分为原有厂房的古老立面，基座之上是13层高的楼体，整座建筑沿着Ciutat de Granada街道呈棱柱形，与周围风格十分融洽。为了巧妙处理街角，也为了强调建筑的狭长感，建筑师在设计时对平面构图做了扭曲处理。

为了让建筑外观显得协调，建筑师在立面的外面又围起了一层统一的立面，新的立面在遮阳设施部位折叠成鳞片状，将需要保留的旧有立面包围起来，使之与整座建筑协调一致，成为一个整体。新建入口处刻有建筑名字的模板巧妙地融进了双立面中。

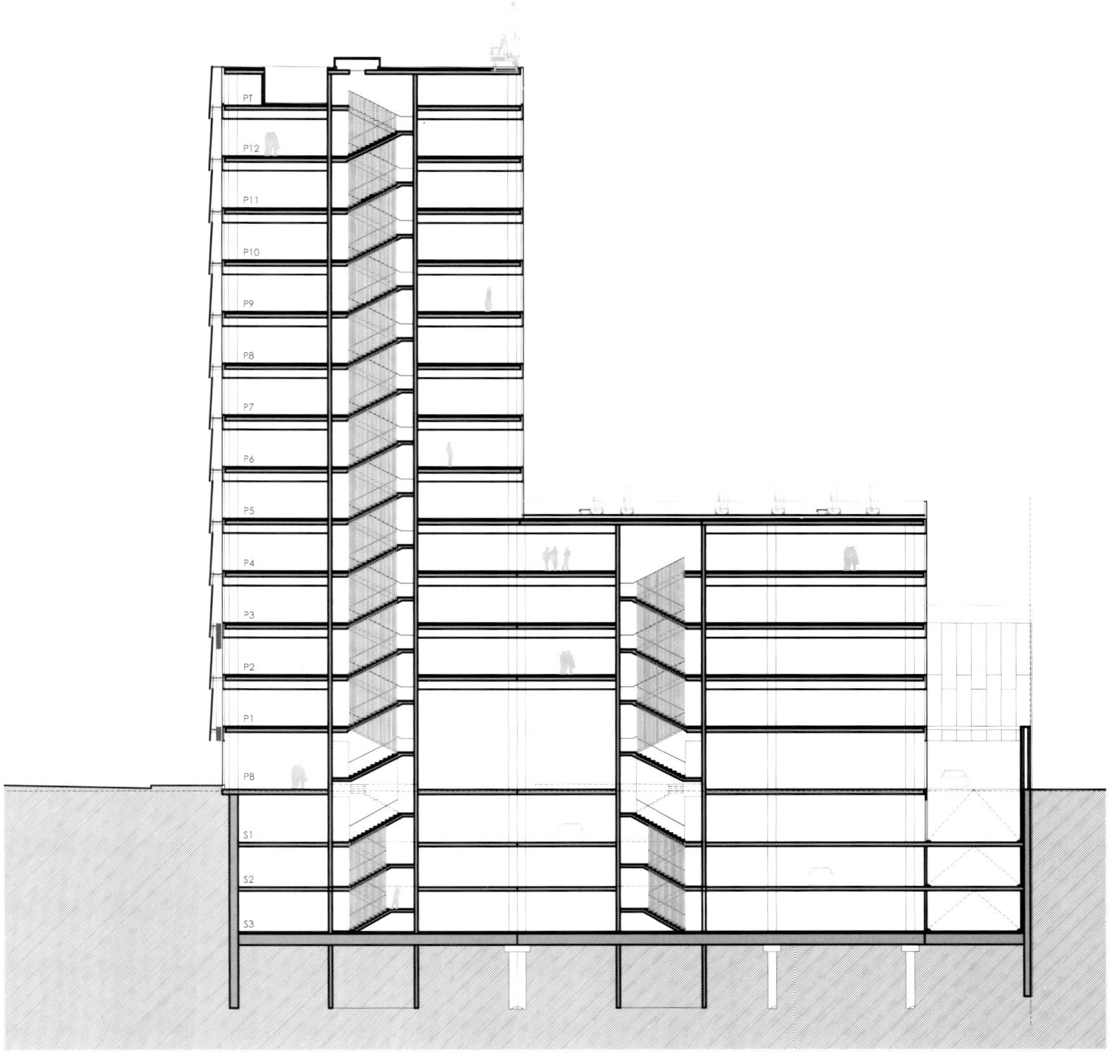

SECCIÓ 1
e: 1/500

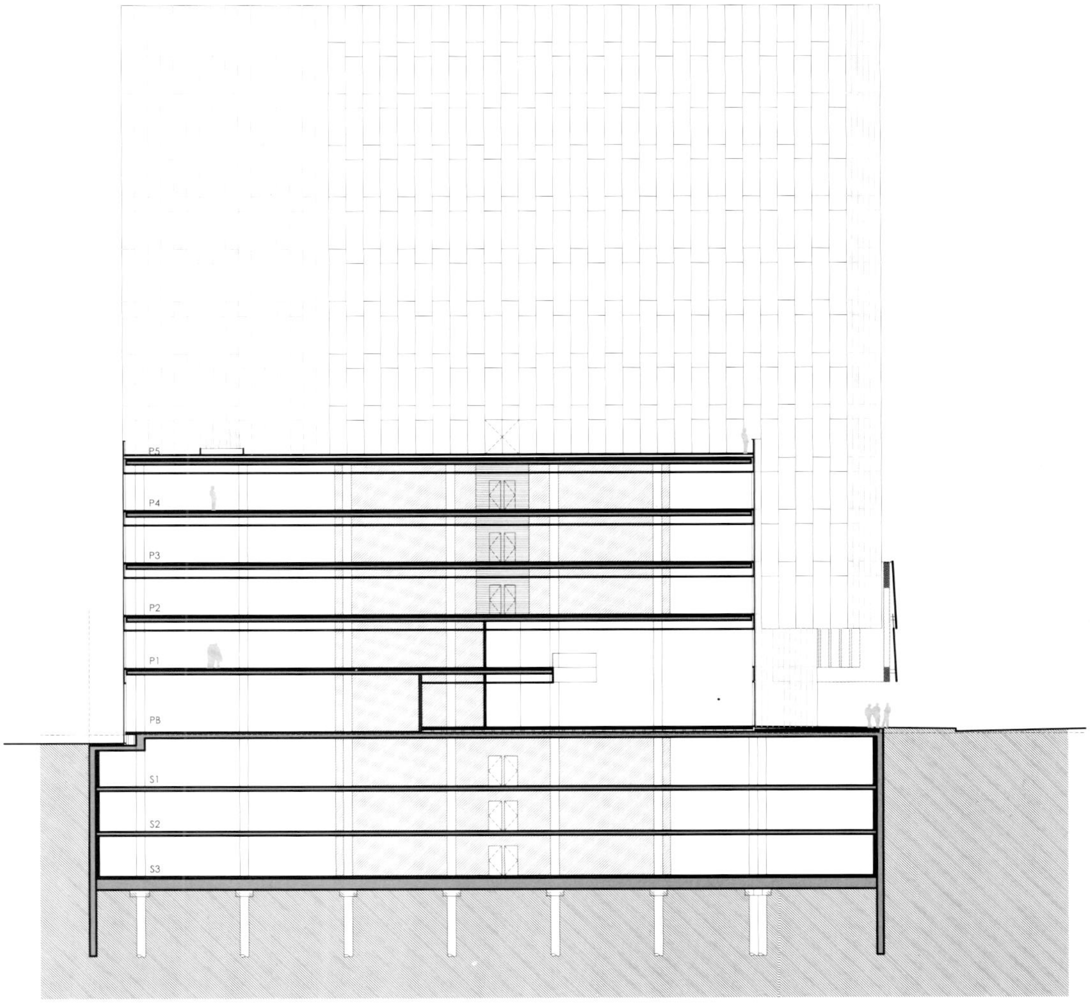

SECCIÓ 2
e: 1/500

1. DOBLE ACRISTALAMIENTO AISLANTE COMPUESTO POR:
   EXTERIOR: LUNA DE 6mm BAJO EMISIVO
   CÁMARA DE AIRE 12mm
   INTERIOR: VIDRIO LAMINAR DE 4+4mm TRANSPARENTE
2. DOBLE ACRISTALAMIENTO AISLANTE COMPUESTO POR:
   EXTERIOR: LUNA DE 6mm CONTROL SOLAR
   CÁMARA DE AIRE 12mm
   INTERIOR: VIDRIO LAMINAR DE 4+4mm TRANSPARENTE
3. LUNA DE 6mm TEMPLADA Y SERIGRAFIADA
4. PERFIL DE ALUMINIO ANODIZADO
5. ANCLAJE DE ACERO GALVANIZADO PROVISTO DE REGULACIÓN TRIDIMENSIONAL
6. PANEL SANDWICH METÁLICO 1mm+30mm (LANA DE ROCA) + 1mm
7. AISLANTE DE LANA DE ROCA
8. PROMATECT-L e: 40mm
9. REJILLA DE ACERO GALVANIZADO EN CALIENTE
10. PIEZA SUJECIÓN VIDRIO DE ACERO INOXIDABLE
11. VIDRIO LAMINAR 2FLOAT DE 10mm TEMPLADOS Y UNIDOS CON BUTILOS Y SERIGRAFÍA
12. MÉNSULA DE ACERO GALVANIZADO
13. DIFUSOR LINEAL DE IMPULSIÓN DE AIRE
14. FALSO TECHO METÁLICO MICROPERFORADO
15. PAVIMENTO TÉCNICO
16. FORJADO DE HORMIGÓN e: 28cm
17. SELLADO DE SILICONA

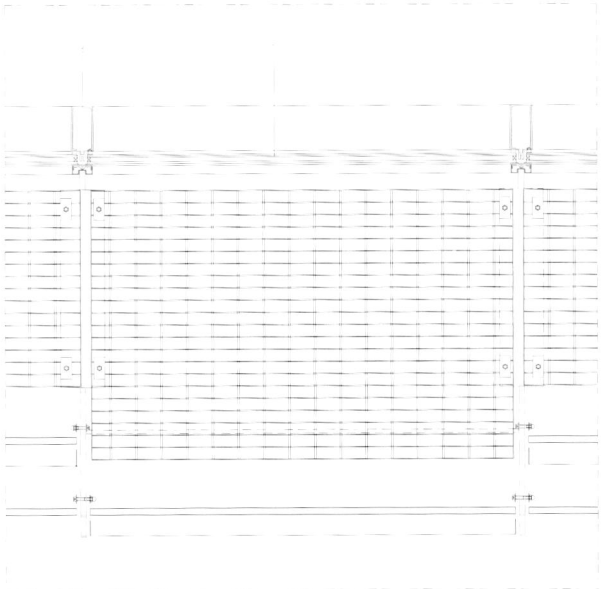

DETALLE FACHADA

IL3

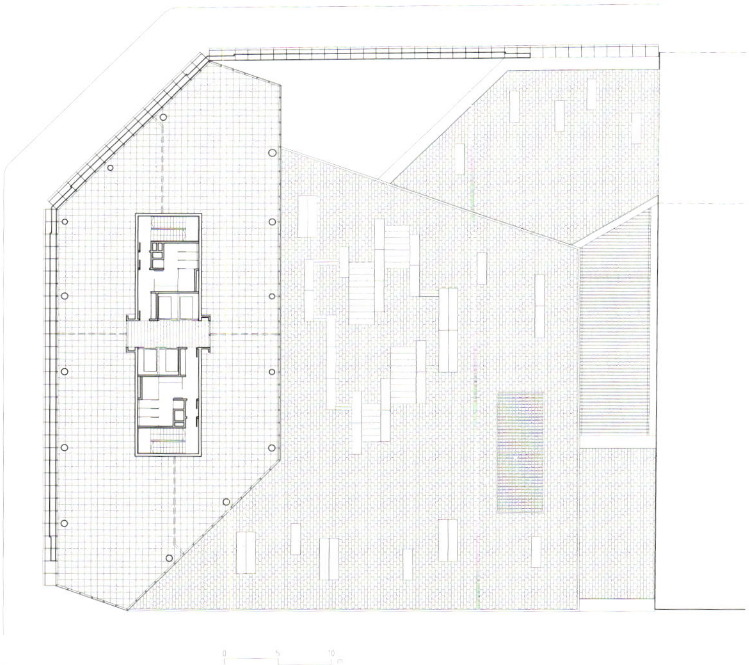

PLANTA SEXTA - DOCEAVA

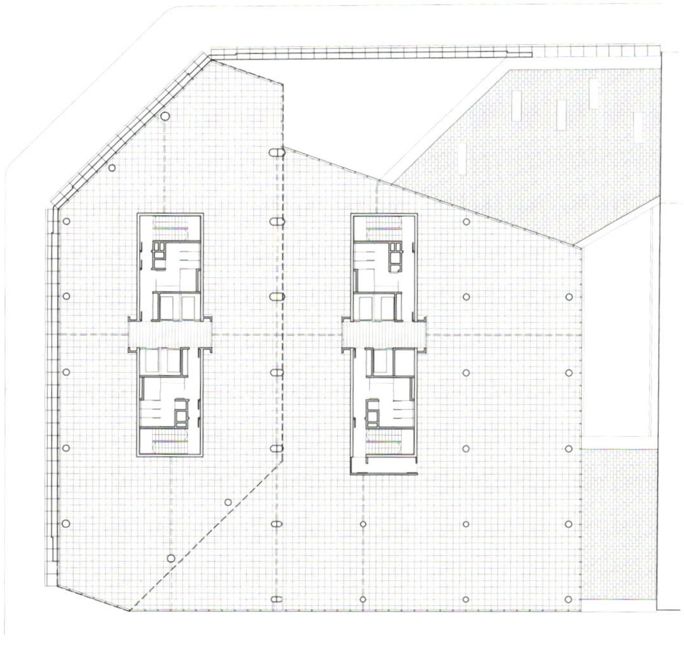

PLANTA CUARTA

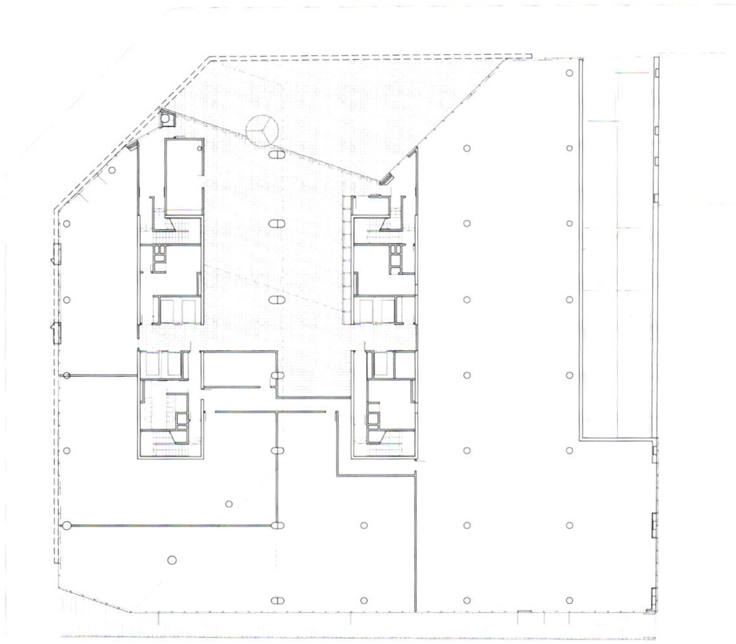

PLANTA BAJA

PLANTA SÓTANO

Fukoku Tower

LOCATION\2-4, Komatsubara, Kita-ku, Osaka, Japan

This tower project for the Fukoku insurance company takes inspiration from the profile of a gigantic tree whose roots proliferate on the surface of the ground. Splayed at its base, the tower's outline tapers elegantly as it rises, gracing the city's skyline with a vertical asymptote. The contrast between the structure's base and upper regions is accentuated by the treatment of the building's "bark". Broad "wood chips" on the lower levels gradually give way to a sleek wall. The glass facade is worked into a crescendo of encrusted mirrors at the base, reflecting the colors of the sky and the surrounding environment. Situated at the exit of Osaka's main train station, the building will become a lasting landmark in the urban landscape.

ARCHITECT
Dominique Perrault

FIRM
Dominique Perrault Architecture, Paris, France

ASSOCIATED ARCHITECT
Shimizu Corporation Architects & Engineers

ENGINEERING
Shimizu Corporation Architects & Engineers

DEVELOPMENT
Development Director: Mitsubishi Estate Group

AREA
68500 m^2

カラオケ

"BIG TREE"

FUKOKU TOWER

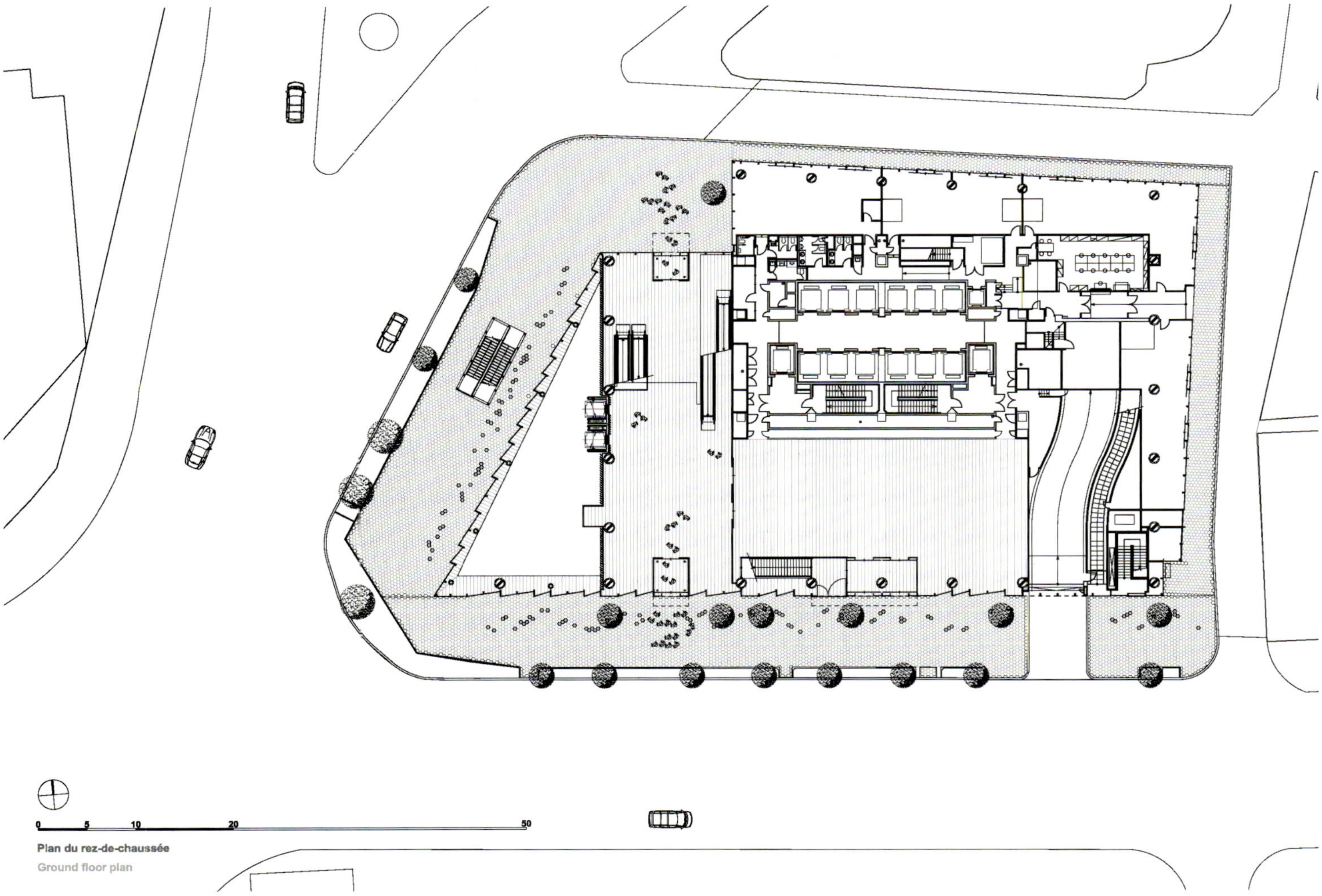

Plan du rez-de-chaussée

Ground floor plan

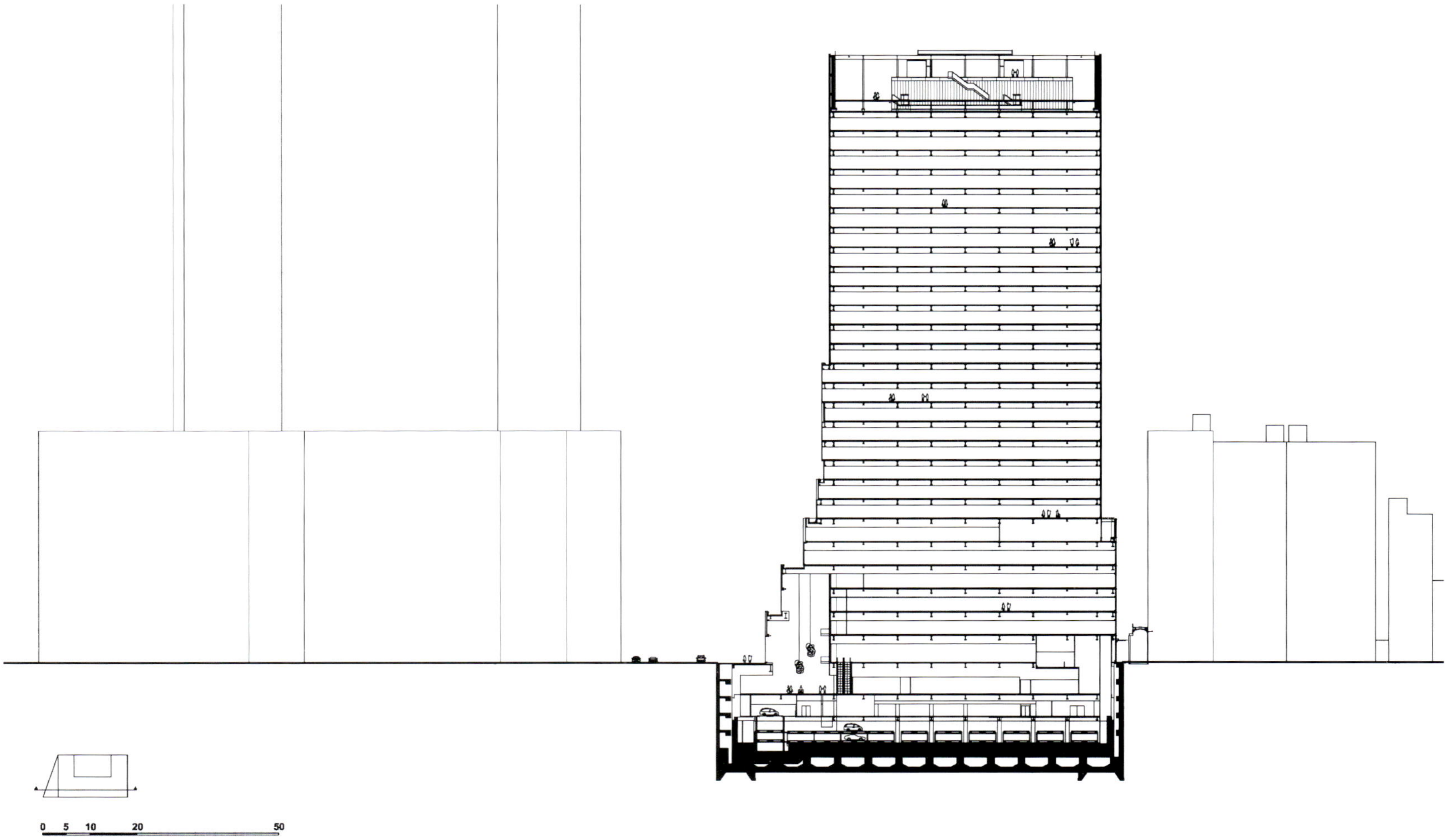

Coupe

Section

本案是为富国保险公司建造的项目，它的设计灵感来源于一棵巨树的形象，这棵大树的根部一直生长到了地面之上。大厦从地基部分展开，如同一棵大树拔地而起，随着高度的升高而愈见纤细，外观尽显优雅，直抵天际线。通过对大厦“树皮”的处理，使大厦基础部分与上半部分形成鲜明的对比。底层宽大的树皮状纹理随着楼层的升高逐渐被光滑的外墙所取代。基座部分的玻璃幕墙如同一个反光效果渐强的带有外壳的镜子，映射着蔚蓝的天空与周边的景物。本案位于大阪市主火车站的出口处，将成为大阪市一道永久的亮丽风景线。

Sanlitun SOHO

By creating a scene where several towers are assembled in a group, the architect wanted to contradict the conventional image of skyscrapers. More specifically, by having 9 similar towers 100 mteres tall standing beside each other, a scene with a "village" of skyscrapers appeared.

With conventional skyscrapers, the objective of each building (an individual or a corporation) was to appeal its own characteristic by its enormous height and shape. However, this "Village of Towers" is not an appeal of one. By sharing a common and soft rule among the buildings, the architect aimed to create a harmonious and peaceful condition.

In order to do so, each tower had to be similar yet different. To create this condition, the team examined the possibilities of a mosaic style design for the façade. The design which has appeared could be called an Asian-style peace. Unlike the western leadership where everybody is subject to a powerful and unique leader, it is the existence of a peaceful and generous group. In the openings between the buildings is a continuous valley-styled space. Water flows in the valleys, which has the potential to become an incubator of informal and wet activities, contrary to the western formal plazas.

ARCHITECT
Kengo Kuma
FIRM
Kengo Kuma & Associates
AREA
465680 m^2

LOCATION\Gongti North Road, Chaoyang District, Beijing, China

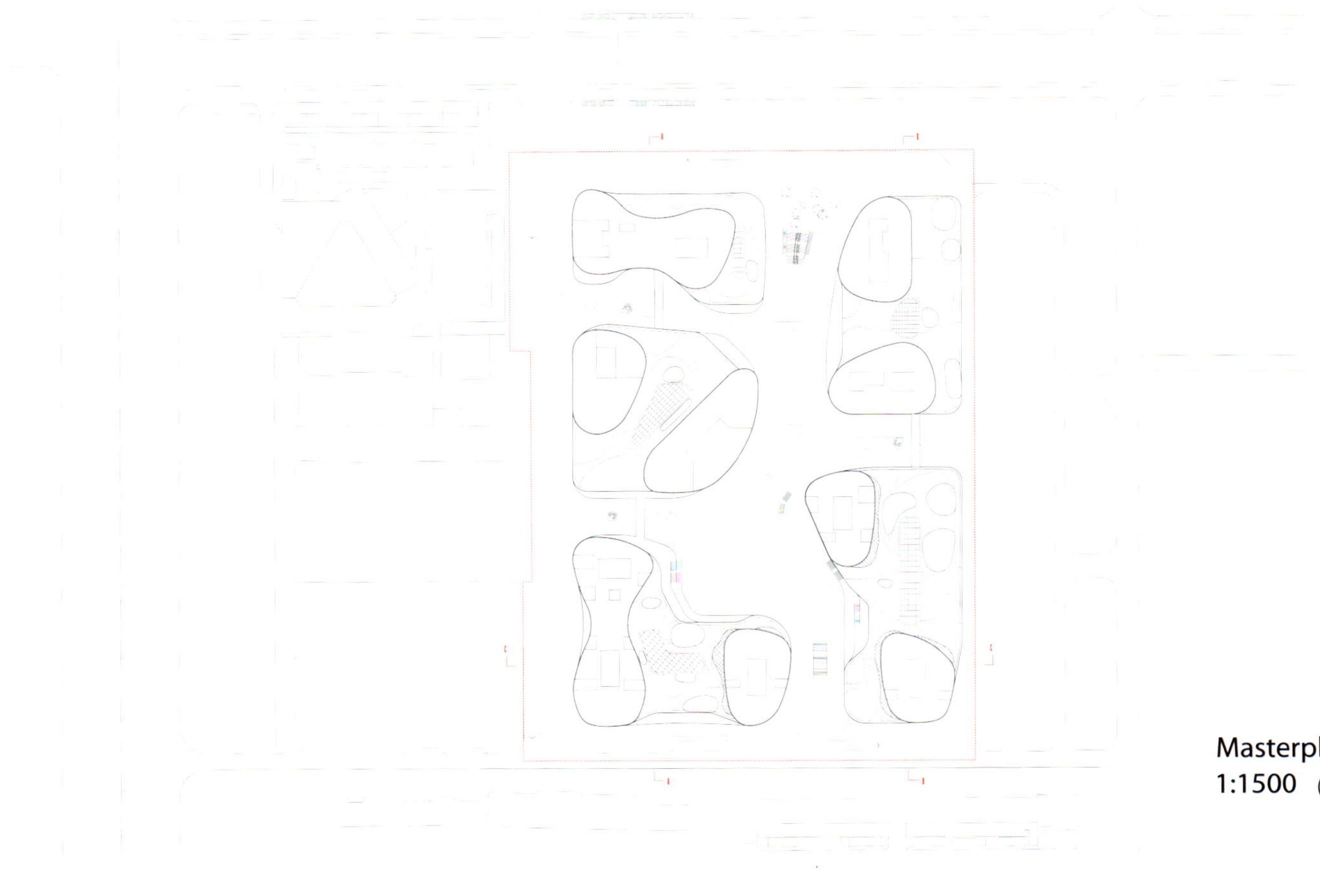

Masterplan
1:1500

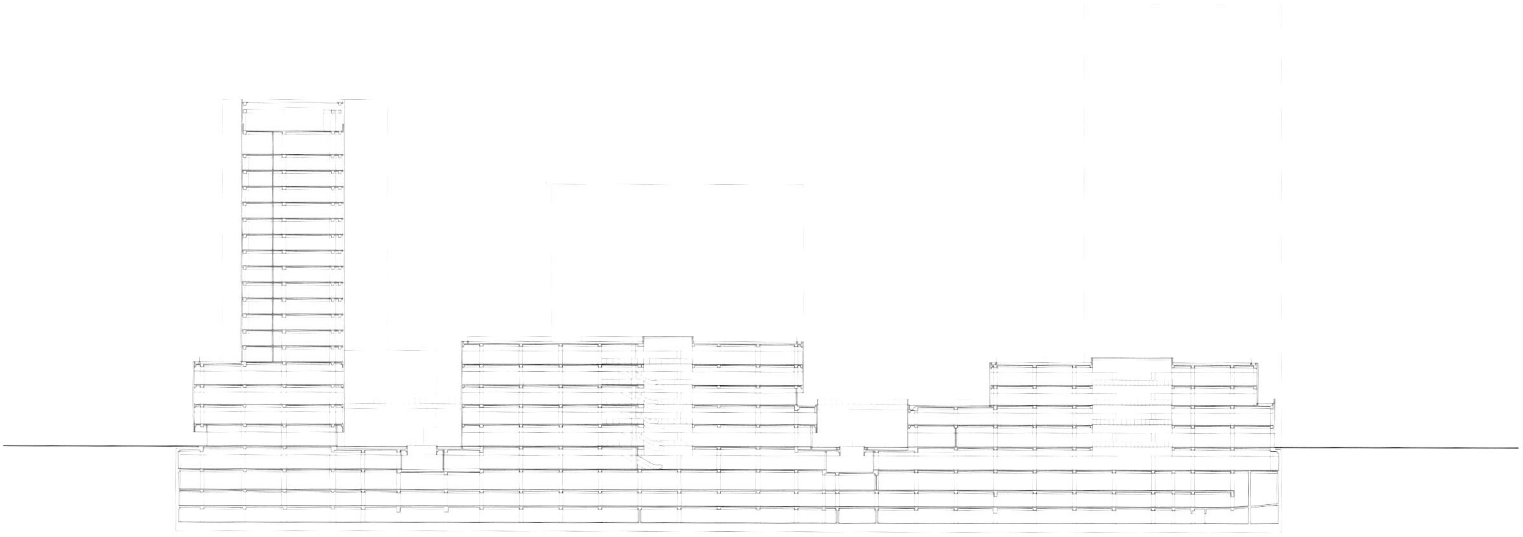

section A 1:800

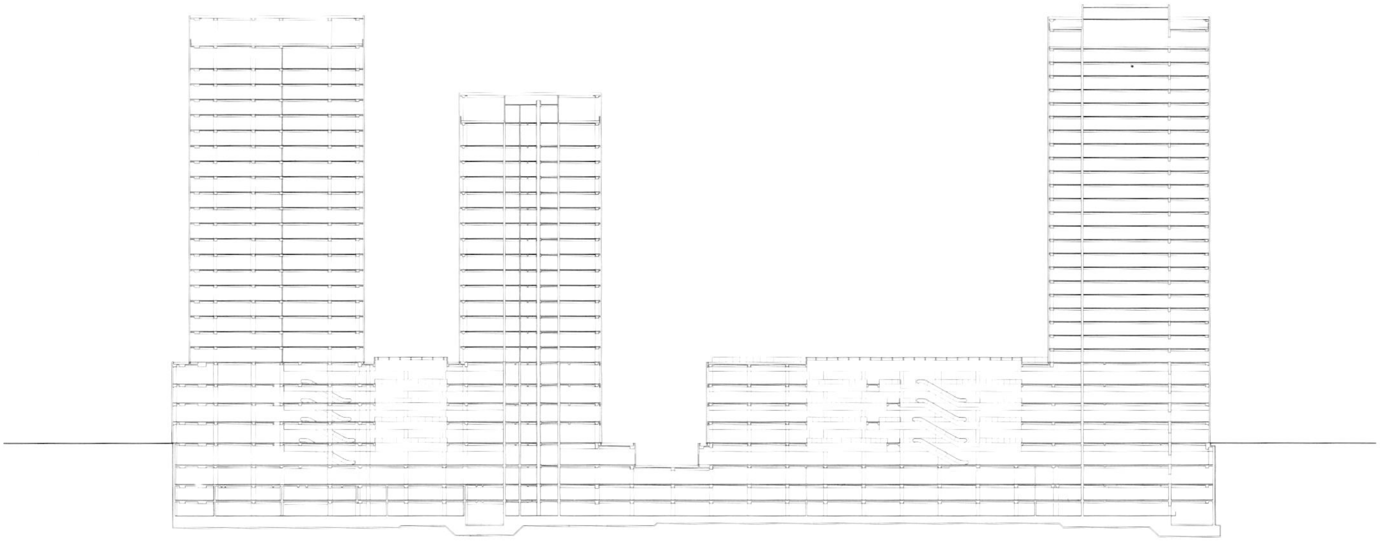

section B 1:800

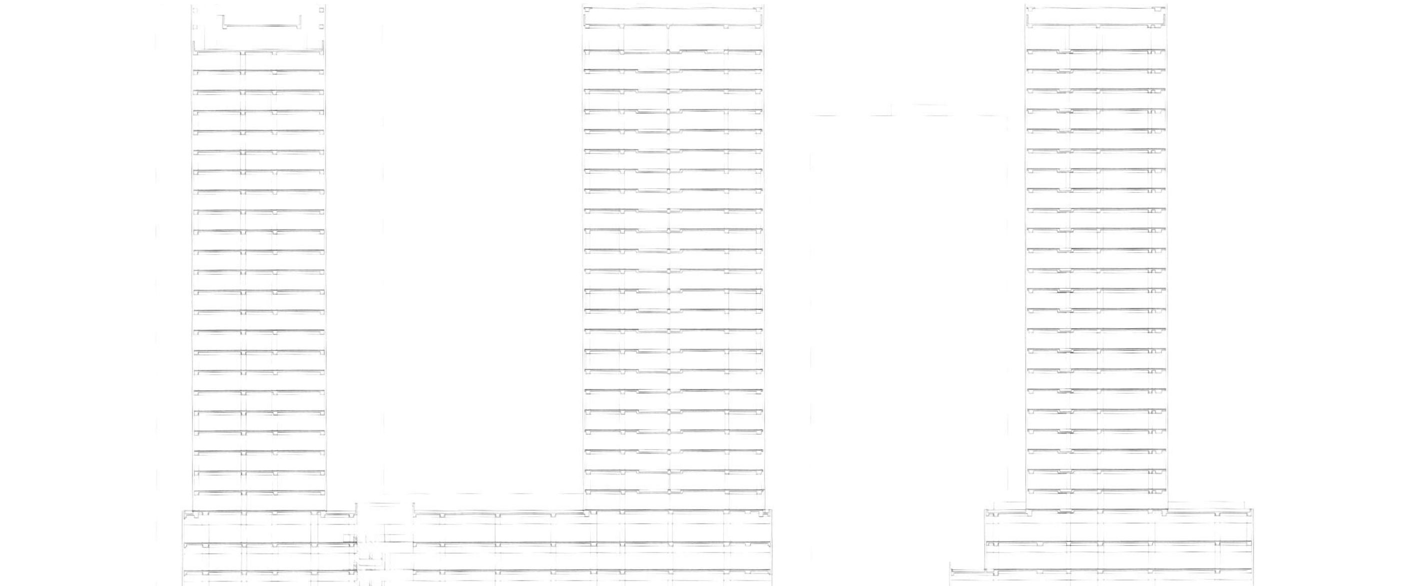

section C 1:800

1:1500

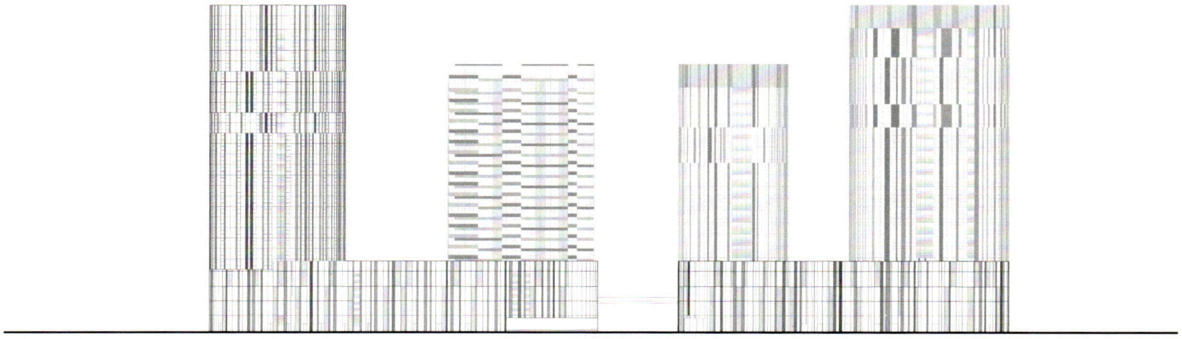

east elevation

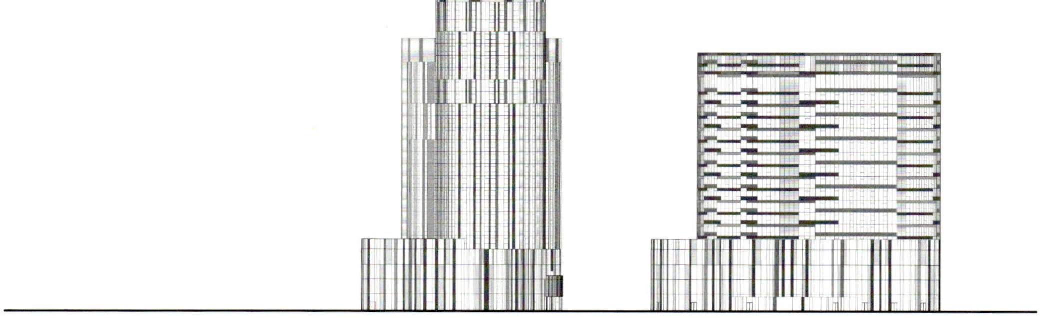

north elevation

1:1500

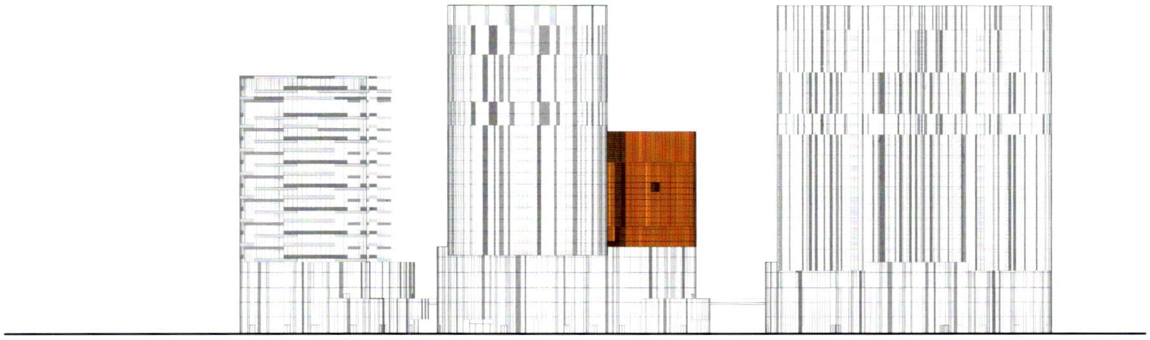

west elevation

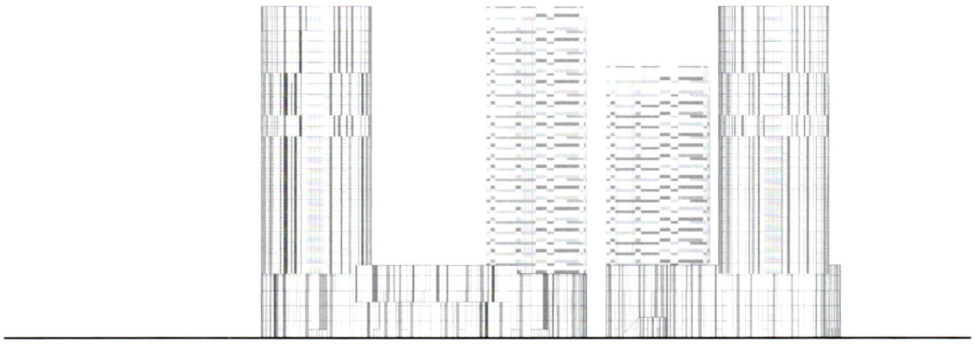

south elevation

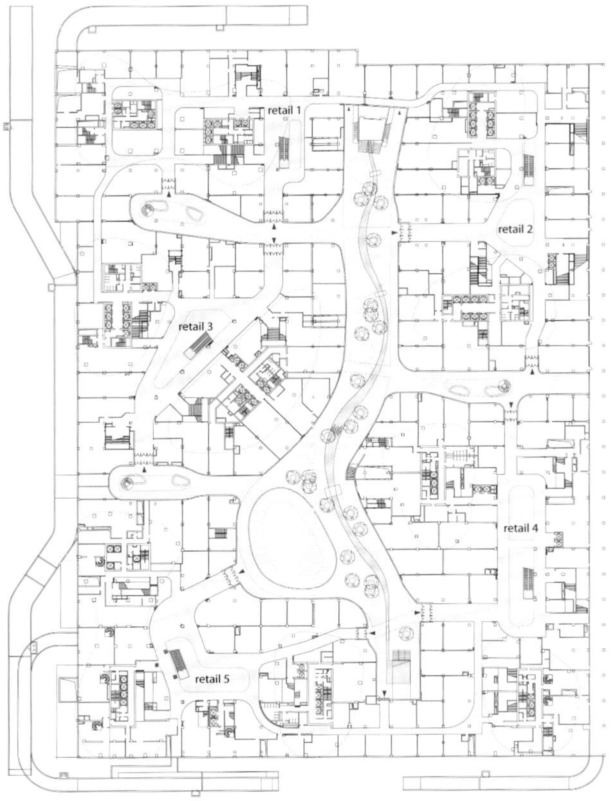

B1F plan 1:800

► retail public entrance

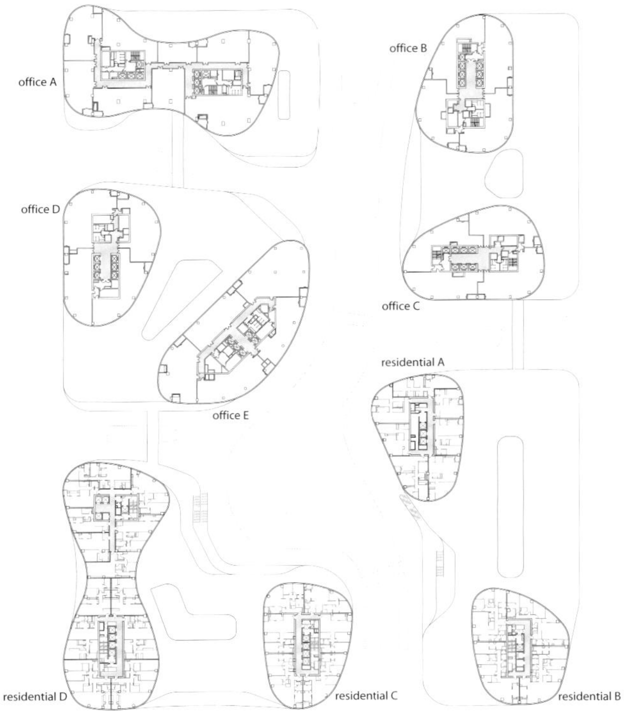

office/ residential tower typical floor plan 1:800

(above F7)

F1 plan 1:800

► retail public entrance
► office lobby entrance
► residential lobby entrance

F3 plan 1:800

► retail public entrance

为了区别于传统的摩天大楼形象，建筑师想通过建一个高塔式建筑群来表现。更确切地说，是将 9 个 100 米高的大楼排列在一起，形成一个摩天大楼“村落”。

与传统的摩天大楼一样，每栋建筑（个人或公司）以其高度和形态展现自身特点。然而，这种“摩天楼村落”不只有一种吸引力，目的是创造一个和谐平静的环境。

为了达到这一点，每座大楼必须既相似又不同。为了验证这种可能性，设计团队使用马赛克材质的外立面达到这种效果。这种设计可以称为亚洲式和平。不同于西方人人都服从于一个强大的唯一领袖，这个建筑群以一种和平、慷慨的团队的形式存在。建筑物之间的开阔地带被打造成连续的类似于山谷的空间。水在山谷中流动，成为水中活动及其他非正式活动的场所，与西方式的整齐的广场正好相反。

U15 Office Building

LOCATION\Assago, Milan, Italy

The disposition on the ground and the typological layout of the U15 office building articulate and specify the general master plan guidelines in relationship to the specific features of the site and of the overall design of the open spaces. The large dimension of the building plot and of the maximum building envelope given by the master plan is interpreted by folding its perimeter inwards to generate a new layout which unites the advantages of an H-shaped scheme to that of a centralized one.

The continuous perimeter "skin" which delimits the interior spaces is divided in several layers which donates depth and a strong light and shadow effect: the inner wall alternates a ribbon window band with an opaque one faced by anodized aluminum panels in natural warm grey, pale gold, light brown shades.

The exterior system of brise-soleils in aluminum sheets folded in different profiles, pierced by an array of custom-designed holes and treated with different anodized natural colors protects the windows from the east and west low sun transforming the building in a sort of large tree-trunk with an iridescent bark.

The design of the surrounding open spaces alternates paved and green surfaces; the former shape the paths reaching the main atrium from south and east, and offering themselves as a pleasant dehors for the ground floor offices.

ARCHITECT
Cino Zucchi Architetti
CLIENT
Milanofiori 2000
AREA
10471 m^2

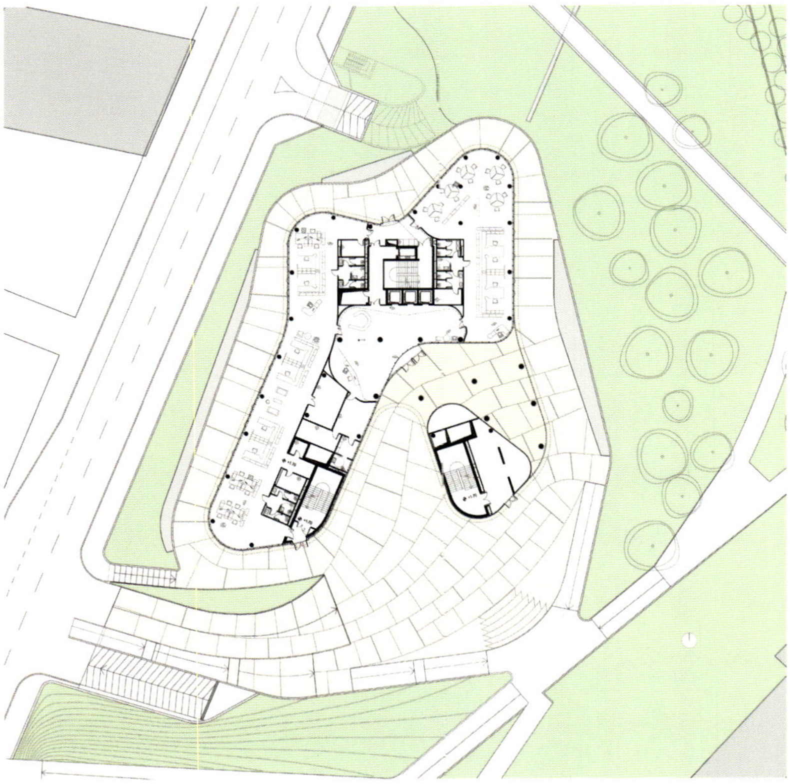

U15 办公楼地面上的配置和典型的布局清晰地表达了其总体设计规划指导方针：即综合考虑大楼所在地的具体特色以及开放空间的整体设计。该大楼总体规划所倡导的建筑理念为建筑用地规模化及使建筑外围结构最大化。这种设计理念是通过其外围向内折叠以产生新的布局空间以及大楼设计成“H”形和集中化设计来体现的。

界定室内空间的绵长的外立面被分为数个层面，以产生纵深感和强烈的光影效果。内墙设计成带状窗户与不透明窗户交替出现的形式，不透明窗户表面所使用的材质是呈自然暖灰、淡黄、浅棕色色调的阳极氧化铝面板。

铝片制成的百叶窗作为建筑物外部系统的遮阳装置，被折叠成不同的形状，中间打有一排专门设计的孔，表面镀上经阳极氧化处理过的各种各样的天然色。这种设计能防止在朝阳或者落日照射下而使建筑物出现斑驳的阴影。

大楼四周的开放空间设计成道路和绿色植物相间的形式。道路从南和东两个方向直通大楼的中庭。中庭和室外道路自然融为一体，为底层的办公人员提供了一个温馨的室外活动空间。

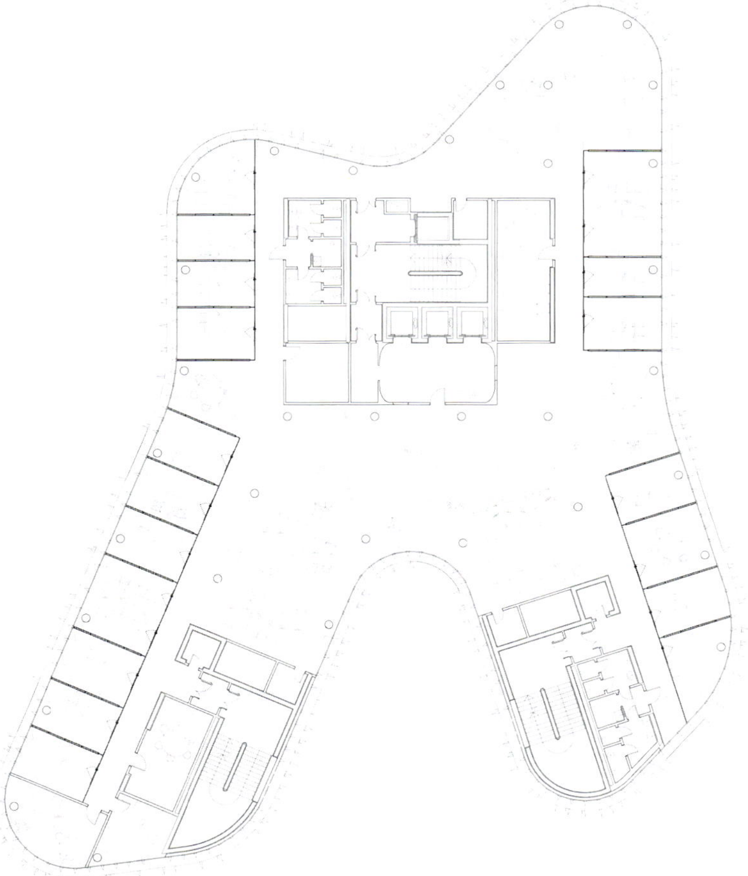

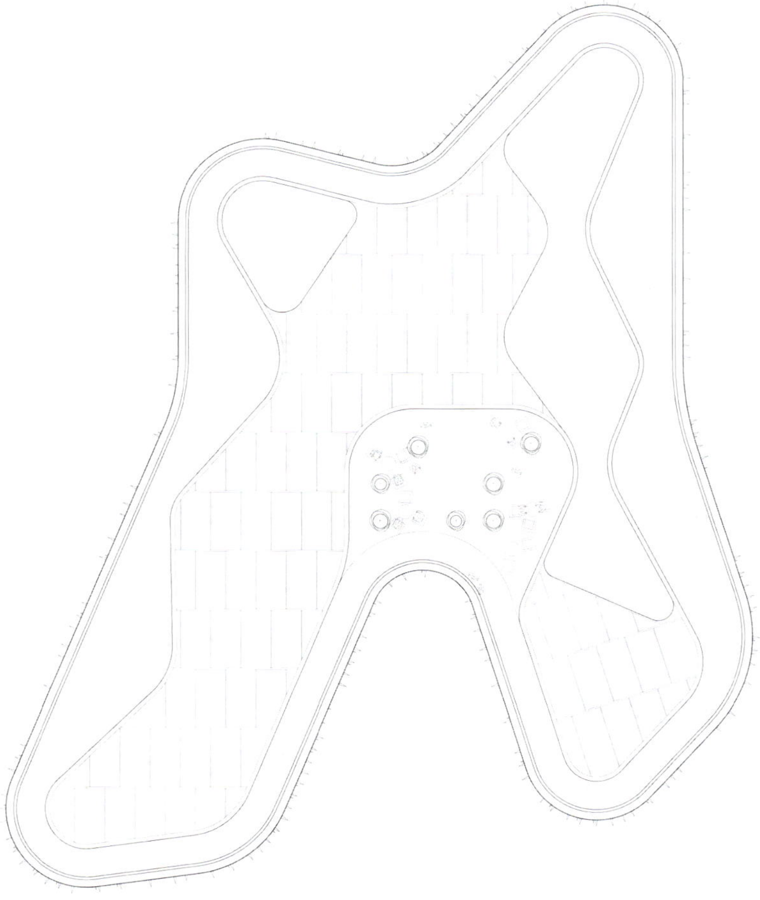

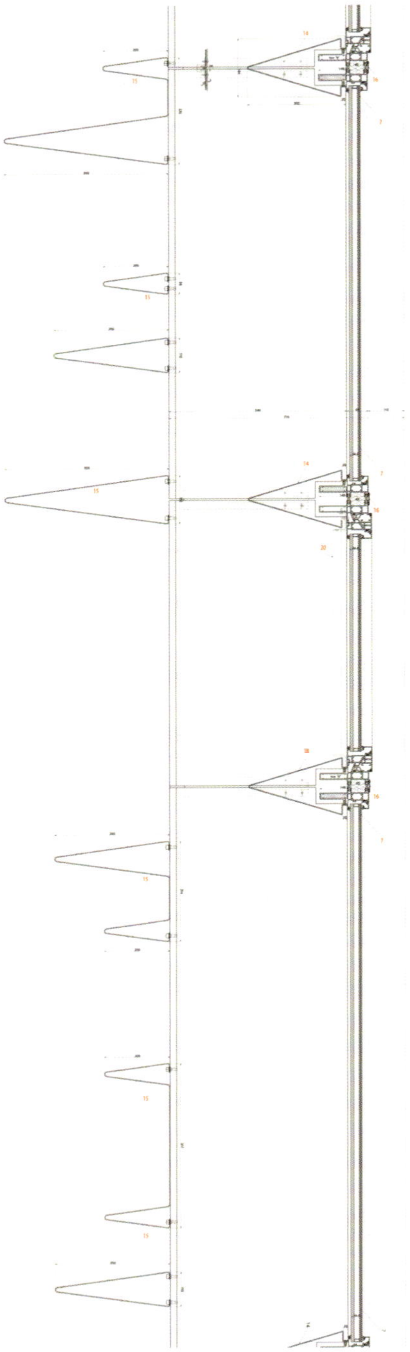

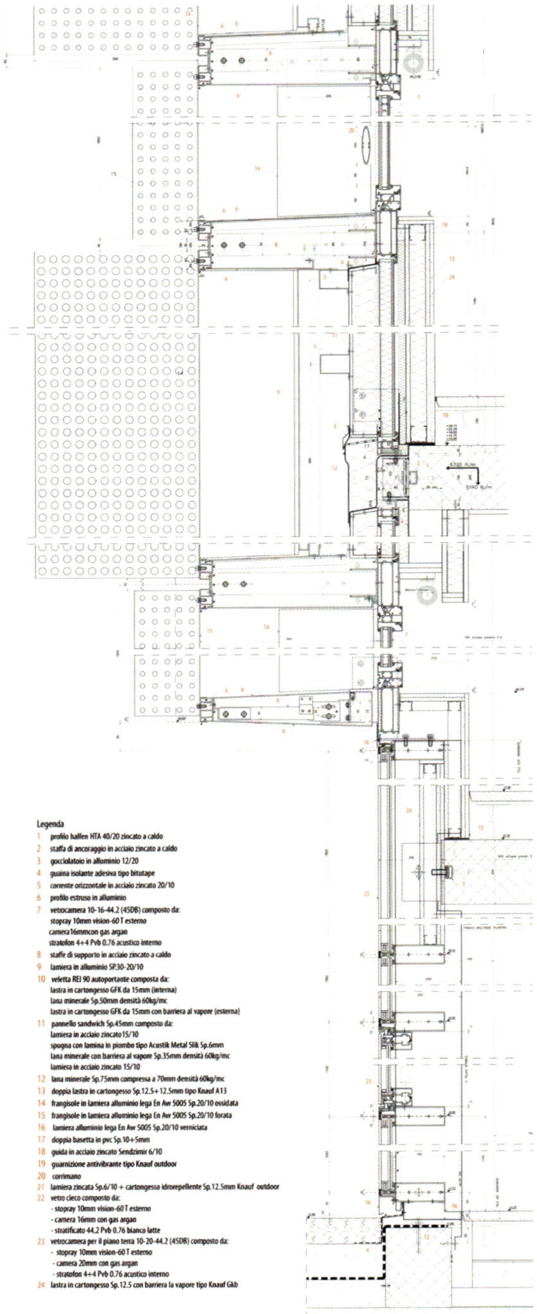

Legenda

1 profilo halfen HTA 40/20 zincato a caldo
2 staffa di ancoraggio in acciaio zincato a caldo
3 gocciolatoio in alluminio 12/20
4 guaina isolante adesiva tipo bitutape
5 corrente orizzontale in acciaio zincato 20/10
6 profilo estruso in alluminio
7 vetrocamera 10-16-44.2 (45DB) composto da:
stopray 10mm vision-60 T esterno
camera16mmcon gas argan
stratofon 4+4 Pvb 0.76 acustico interno
8 staffe di supporto in acciaio zincato a caldo
9 lamiera in alluminio SP.30-20/10
10 veletta REI 90 autoportante composta da:
lastra in cartongesso GFK da 15mm (interna)
lana minerale Sp.50mm densità 60kg/mc
lastra in cartongesso GFK da 15mm con barriera al vapore (esterna)
11 pannello sandwich Sp.45mm composto da:
lamiera in acciaio zincato15/10
spugna con lamina in piombo tipo Acustik Metal Slik Sp.6mm
lana minerale con barriera al vapore Sp.35mm densità 60kg/mc
lamiera in acciaio zincato 15/10
12 lana minerale Sp.75mm compressa a 70mm densità 60kg/mc
13 doppia lastra in cartongesso Sp.12.5+12.5mm tipo Knauf A13
14 frangisole in lamiera alluminio lega En Aw 5005 Sp.20/10 ossidata
15 frangisole in lamiera alluminio lega En Aw 5005 Sp.20/10 forata
16 lamiera alluminio lega En Aw 5005 Sp.20/10 verniciata
17 doppia basetta in pvc Sp.10+5mm
18 guida in acciaio zincato Sendzimir 6/10
19 guarnizione antivibrante tipo Knauf outdoor
20 corrimano
21 lamiera zincata Sp.6/10 + cartongesso idrorepellente Sp.12.5mm Knauf outdoor
22 vetro cieco composto da:
- stopray 10mm vision-60 T esterno
- camera 16mm con gas argan
- stratificato 44.2 Pvb 0.76 bianco latte
23 vetrocamera per il piano terra 10-20-44.2 (45DB) composto da:
- stopray 10mm vision-60 T esterno
- camera 20mm con gas argan
- stratofon 4+4 Pvb 0.76 acustico interno
24 lastra in cartongesso Sp.12.5 con barriera la vapore tipo Knauf Gkb

Legenda
1 profilo halfen HTA 40/20 zincato a caldo
2 staffa di ancoraggio in acciaio zincato a caldo
3 gocciolatoio in alluminio 12/20
4 guaina isolante adesiva tipo bitutape
5 corrente orizzontale in acciaio zincato 20/10
6 profilo estruso in alluminio
7 vetrocamera 10-16-44.2 (45DB) composto da:
stopray 10mm vision-60 T esterno
camera16mmcon gas argan
stratofon 4+4 Pvb 0.76 acustico interno
8 staffe di supporto in acciaio zincato a caldo
9 lamiera in alluminio SP.30-20/10
10 veletta REI 90 autoportante composta da:
lastra in cartongesso GFK da 15mm (interna)
lana minerale Sp.50mm densità 60kg/mc
lastra in cartongesso GFK da 15mm con barriera al vapore (esterna)
11 pannello sandwich Sp.45mm composto da:
lamiera in acciaio zincato15/10
spugna con lamina in piombo tipo Acustik Metal Slik Sp.6mm
lana minerale con barriera al vapore Sp.35mm densità 60kg/mc
lamiera in acciaio zincato 15/10
12 lana minerale Sp.75mm compressa a 70mm densità 60kg/mc
13 doppia lastra in cartongesso Sp.12.5+12.5mm tipo Knauf A13
14 frangisole in lamiera alluminio lega En Aw 5005 Sp.20/10 ossidata
15 frangisole in lamiera alluminio lega En Aw 5005 Sp.20/10 forata
16 lamiera alluminio lega En Aw 5005 Sp.20/10 verniciata
17 doppia basetta in pvc Sp.10+5mm
18 guida in acciaio zincato Sendzimir 6/10
19 guarnizione antivibrante tipo Knauf outdoor
20 corrimano
21 lamiera zincata Sp.6/10 + cartongesso idrorepellente Sp.12.5mm Knauf outdoor
22 vetro cieco composto da:
- stopray 10mm vision-60 T esterno
- camera 16mm con gas argan
- stratificato 44.2 Pvb 0.76 bianco latte
23 vetrocamera per il piano terra 10-20-44.2 (45DB) composto da:
- stopray 10mm vision-60 T esterno
- camera 20mm con gas argan
- stratofon 4+4 Pvb 0.76 acustico interno
24 lastra in cartongesso Sp.12.5 con barriera la vapore tipo Knauf Gkb

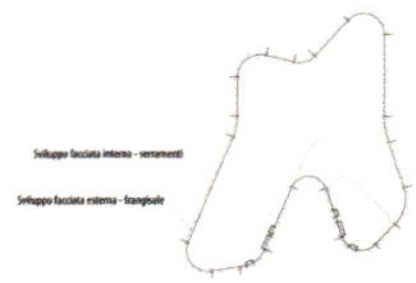

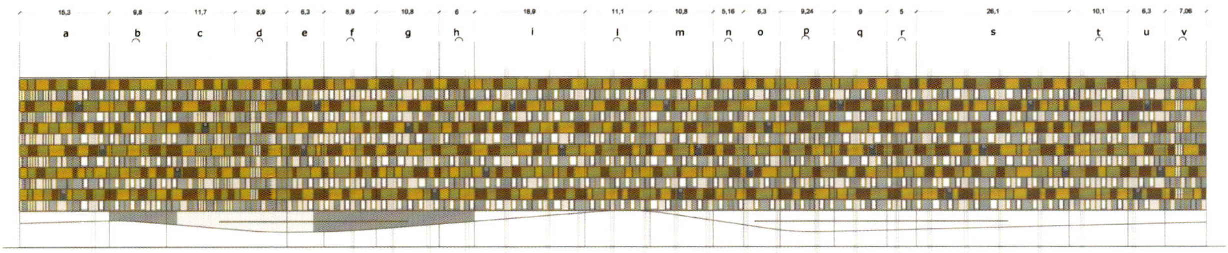

Sviluppo facciata interna - serramenti

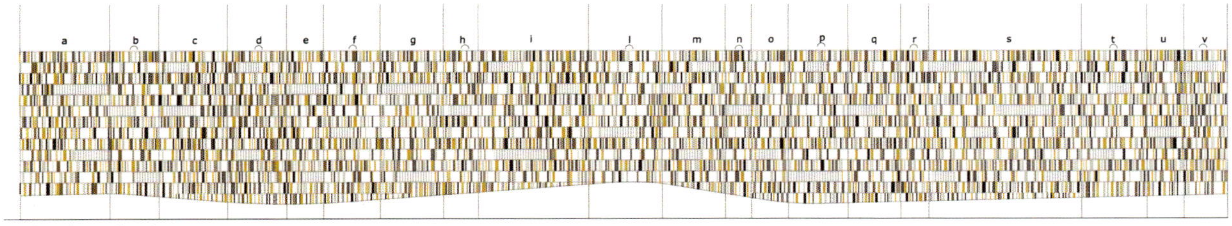

Sviluppo pelle esterna - Frangisole

IT-Fornebu Portal Building

The new Portal building is an extension to the old Airport Terminal building housing the business and incubation centre for information technology companies—IT-Fornebu. The Fornebu area is experiencing a surge in development, transforming a previously busy Airport into a new city-wide destination for business. A-lab was commissioned to develop the building, after winning the competition in 2004.

CRITERIA

In the design A-lab prioritised people. The design criteria's were to reduce volume and waste, provide an effective, efficient and healthy workplace, enhance communications and give a flexible layout.

CONCEPT

The main concept was to strip and divide the program into independent architectonic elements, clearly defined by function, in order to gain maximal flexibility and distinct architectural identity. A continuous base resolves the height difference of the terrain between the main street and the existing terminal building, and forms a multifunctional floor for the Portal building. The project consists of four glazed office blocks, where light, views and acoustics are optimal, four stone clad supply towers and the public hub covered in "mandarin red" aluminum composite panels, lifted over the main entrance.

BUILDING

The four office blocks have a rectangular and rational form and are relieved of all the internal cores and vertical structure allowing an efficient and flexible internal organisation, based on a flexible open floor, a tabula rasa plan with a variety of possibilities. Each office block has 5 floors with a total area of 3,750m^2.

The structure is set outside the exterior façade. It is shaped as a diagonal steel structure around the block and is attached every 4.8 meters with a steel joint through the façade to the floor beams.

All shared functions such as vertical communication, technical services and wet cores are placed in intermediate towers, detached from the working areas. The towers are designed as closed massive monoliths, where the program that is not in need of daylight or views is located.

The free-form orange hub is an exception and has a clear identity, it acts as a strategic element that glues all the activities, highlights the synergy between the existing terminal building and the new IT-Fornebu Portal Building.

ARCHITECT
Jan Petter Seim, Tonje Løvdahl, Tor Inge Hjemdal, Jarand Midtgaard, Inger Totland

FIRM
A-lab

LANDSCAPE DESIGN
Asplan Viak

STRUCTURAL ENGINEER
Rambøll Norge AS (competition phase: Arup)

M&E SERVICE
Electro (Føyn Consult)/ Ventilation (Norconsult)

STRUCTURE
Steel and concrete (prefabricated)

FINISH
Glass, stone and aluminium composite panels

PARTNER
Odd Klev, Geir Haaversen, Adnan Harambasic

AREA
28000 m^2

PHOTOGRAPHER
Luis Fonseca

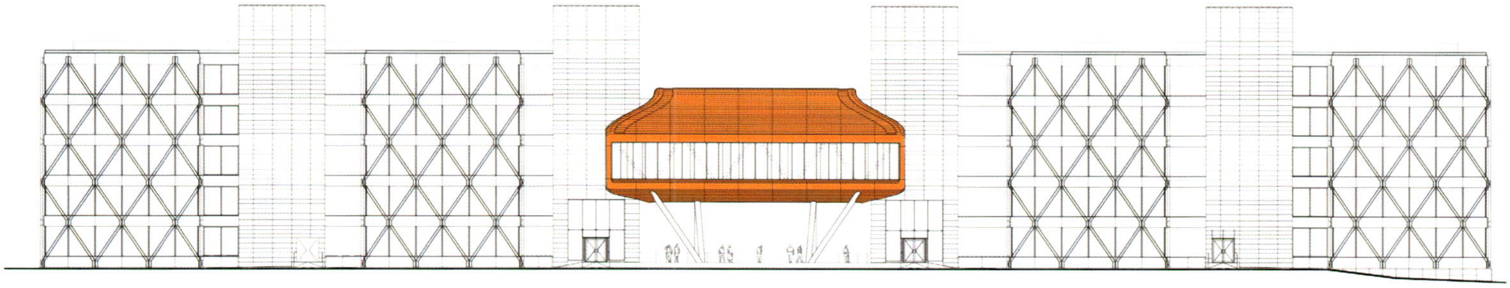

Facade south

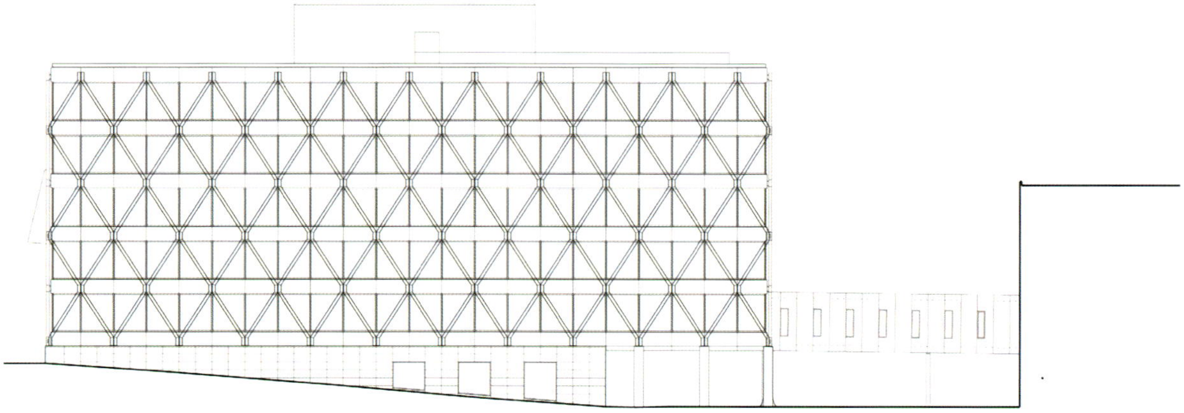

Facade east

vertical detail section

1. exposed external steel truss, ø 219mm
2. fastening, steel, 4800mm o.c.
3. aluminum composite panel
4. Sealed double-glazed window
5. louvers
6. windowframe,190/65 aluminum, coated
7. plaster board 13 mm
8. cableduct
9. radiator
10. insulation 200 mm
11. plywood 18 mm
12. fire insulation 20 mm
13. HSQ
14. concrete hollowcore320 mm
15. system ceiling

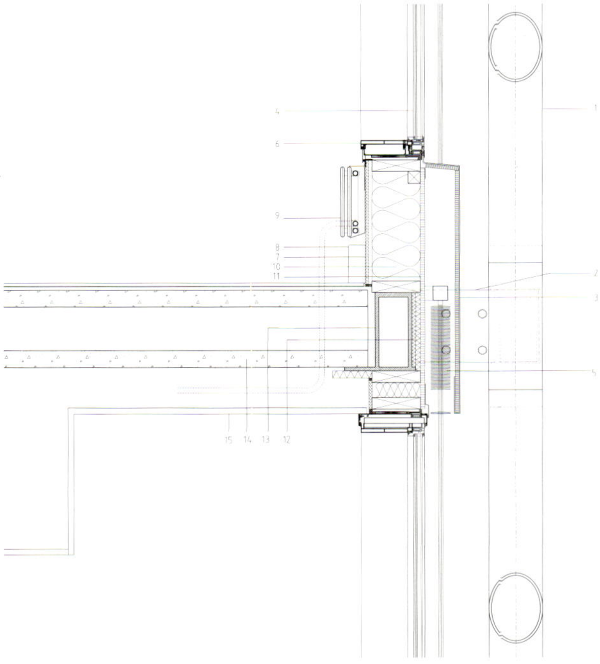

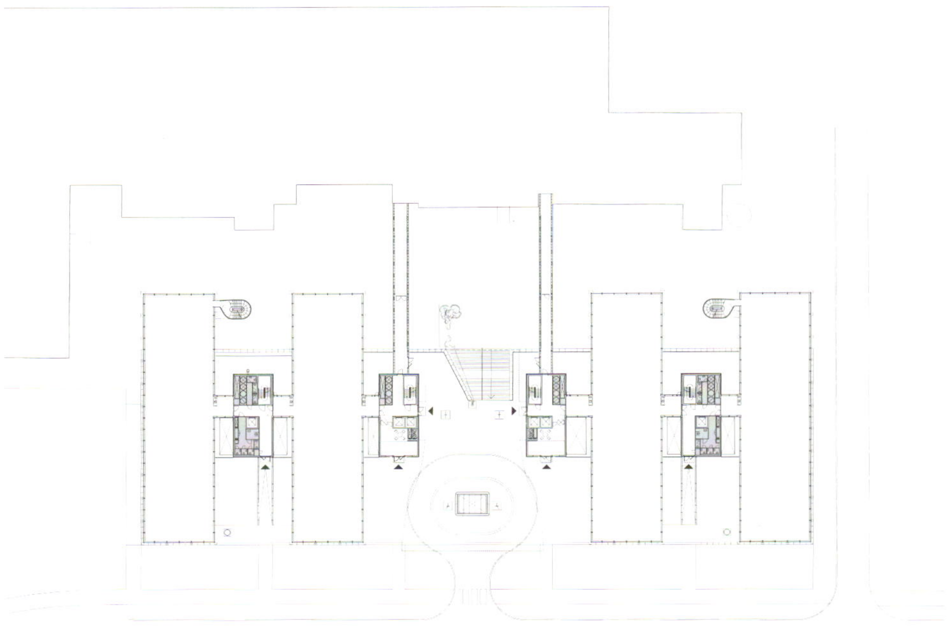

Groundfloor

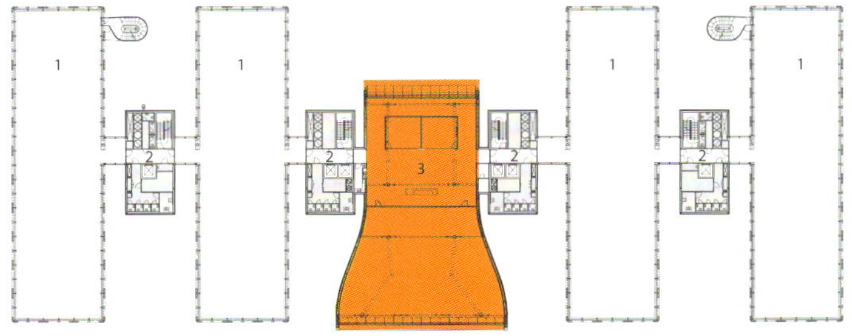

3rd floor

1: OFFICE
2: SUPPLY TOWER
3: MEGAHUB

IT-Fornebu 公司新办公大楼是 IT-Fornebu 公司改造和扩建以前的机场候机楼而建成的，经过脱胎换骨式的改造，先前的机场变成了新的城市商业中心，多家信息公司的商业和服务中心皆驻扎在此。A-lab 建筑设计公司的设计在 2004 年的竞赛中获胜。

标准

A-lab 的设计理念优先考虑人的因素。其设计标准是减少体量和浪费，建造一个实际高效的健康工作环境，增强人们的交流和灵活布局。

理念

该项目的主要设计理念是将整个项目分离成几个独立的建筑单位，每个单位有清晰的功能定位以便于灵活使用，同时每座建筑也有独特的身份。一块完整的地基解决了主要通道和候机楼之间地形高低不一的难题。整个项目包含四座装有玻璃幕墙的办公大楼，这四座大楼里的灯光设施、视野和音响都达到最优化。项目还有四座石材幕墙的供应楼和一个铝复合板包面的橘红色公共中心，橘红色公共中心下方就是该设计的主入口。

四座办公大楼呈矩形，因此大楼内部中心区域就处于开放状态。垂直结构也使建筑内部更高效、灵活。各楼层布局灵活，开放的空间对使用者来说就是可进行各种设计的“白板”。每个高五层的办公楼面积为 3750 平方米。

在办公楼外立面的表面有呈对角线交叉造型的钢梁结构，每根长 4.8 米的钢梁与楼体的楼层梁相连。

一些公共功能单位，如纵向沟通区域、技术服务区以及厨房、厕所、浴室等设置在中间的供应楼上。供应楼与办公楼呈分离状态，对自然光和景观要求不高的项目被安排在供应楼中。

在整个设计中，成自由形态的橘红色的公共中心超然于整个建筑群之外，在整个建筑中有着清晰的定位，这里是将所有活动融合起来的元素，特别突出了先前的候机楼和新办公大楼之间的和谐。

ventelo
Software Innovation
Ringerike Liftutleie AS

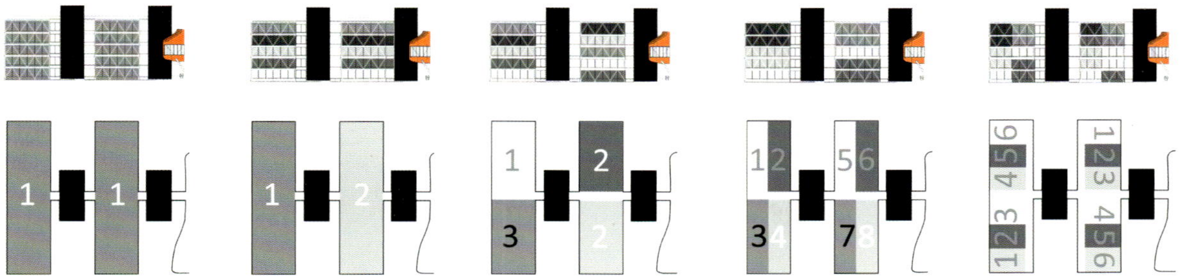

	Burolandschaft offices	Traditional Britsh speculative offices	New «Broadgate» type of British speculative office	Traditional North American speculative office	North European office	It-Fornebu Portalbuilding
typological efficiency						
No. of storeys	5	10	10	80	5	5
Typical floor size	$2000m^2$	$1000m^2$	$3000m^2$	$3000m^2$	$2000m^2$	$800m^2$ x 4
Typical floor depth	40m	13.5m	18m and 12m	18m	10m	50m
Furthest distance from perimeter aspect	20m	7m	9-12m	18m	5m	14m
Efficiency: net to gross		80%	85%	90%	70%	0%
Maximum cellularization (%of usable)	20%	70%	40%	20%	80%	100%
Type of core	semi-dispersed	semi-dispersed	Concentrated extremely compact	Concentrated extremely compact	Dispersed: stairs more prominent than lifts	Concentrated external

sourse: Duffy, «the responsible workplace»

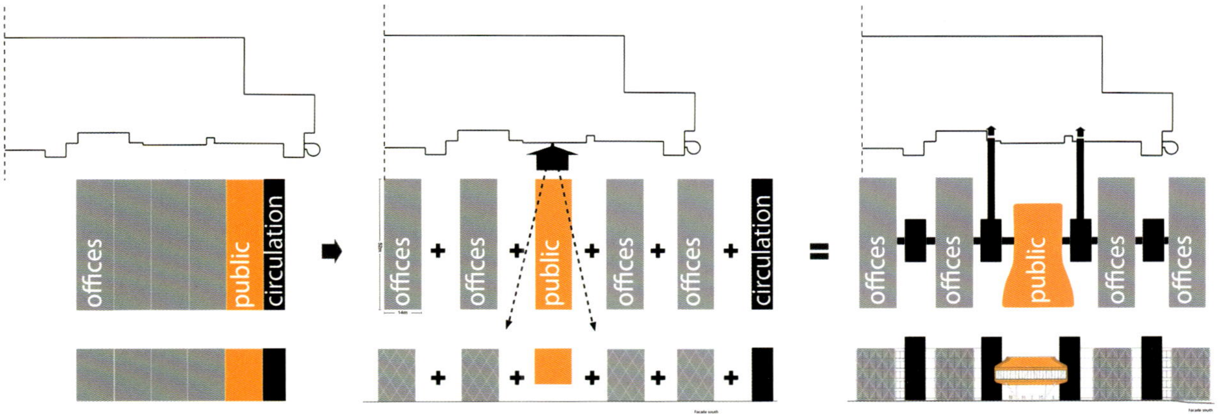

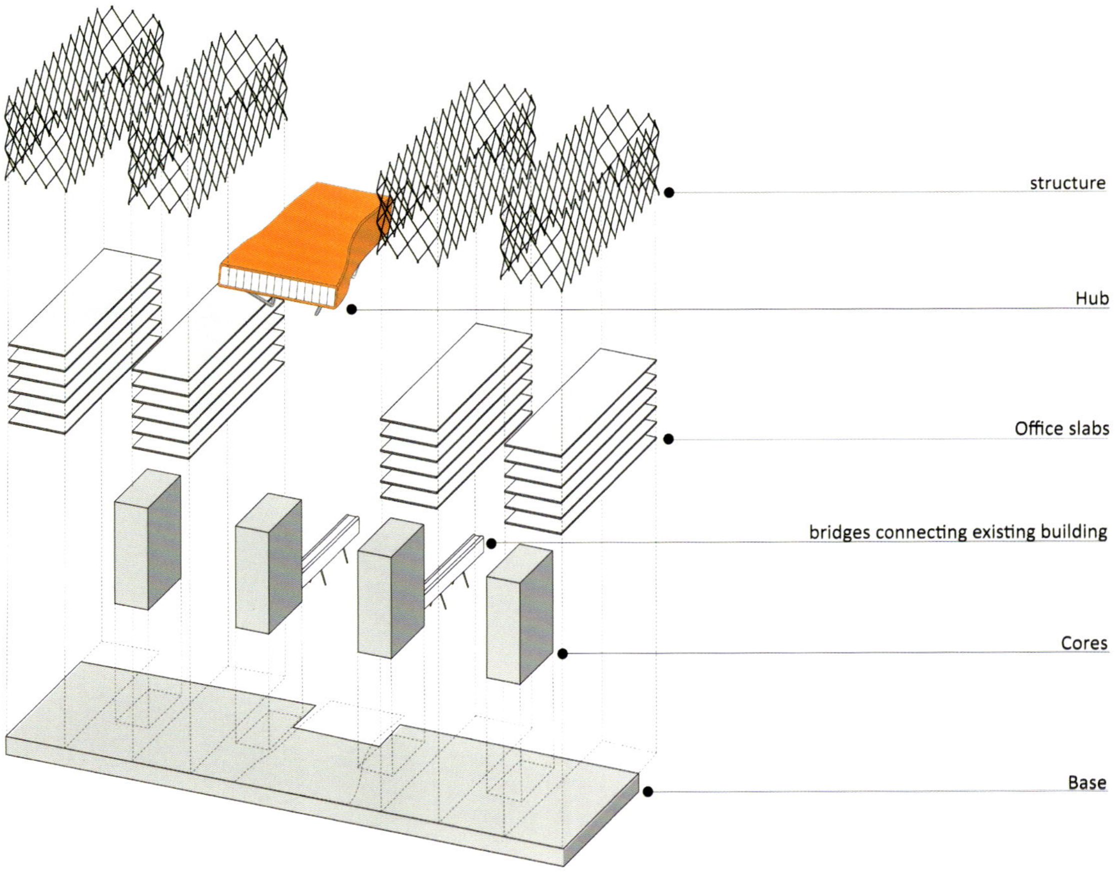

ITFORNEBU

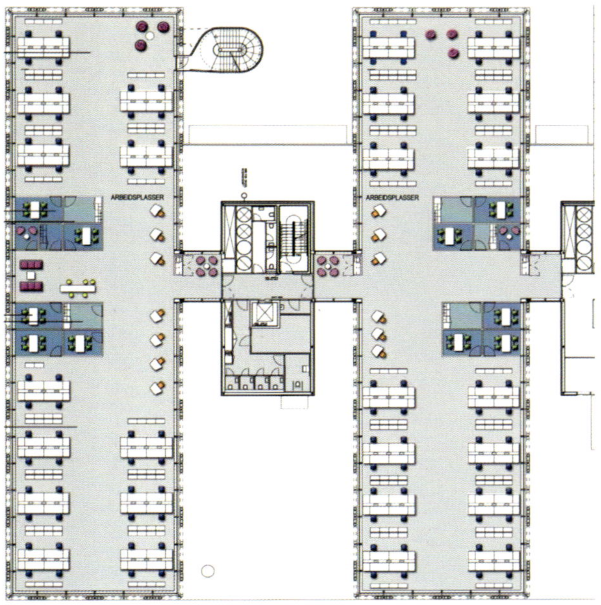
ARBEIDSPLASSER
ARBEIDSPLASSER

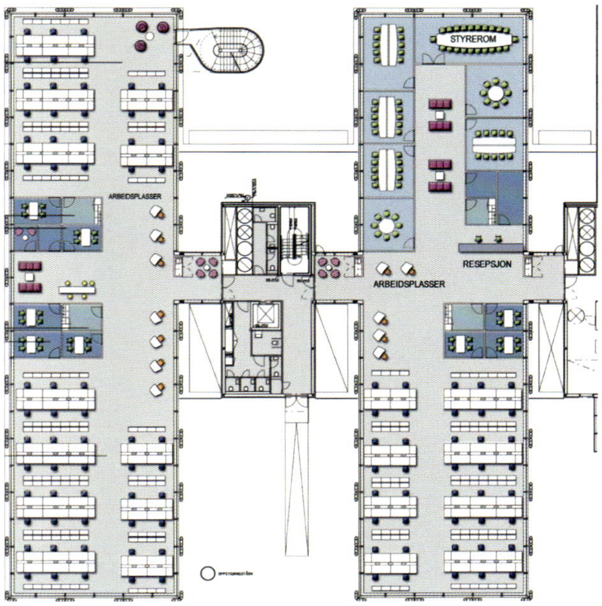
STYREROM
ARBEIDSPLASSER
RESEPSJON
ARBEIDSPLASSER

OTHER FUNC
(MARKET ACC
ADMIN.)
POSTAL ROOM
RECEPTION
PUBLIC SPACE
SMALL KITCHEN
SMALL KITCHEN
PUBLIC SPACE
WARDROBE
STORAGE
ARCHIVE

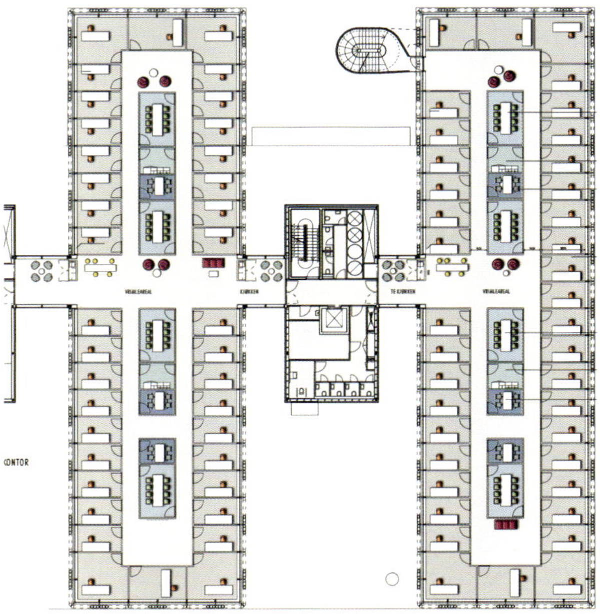

International Management Institute (IMI)

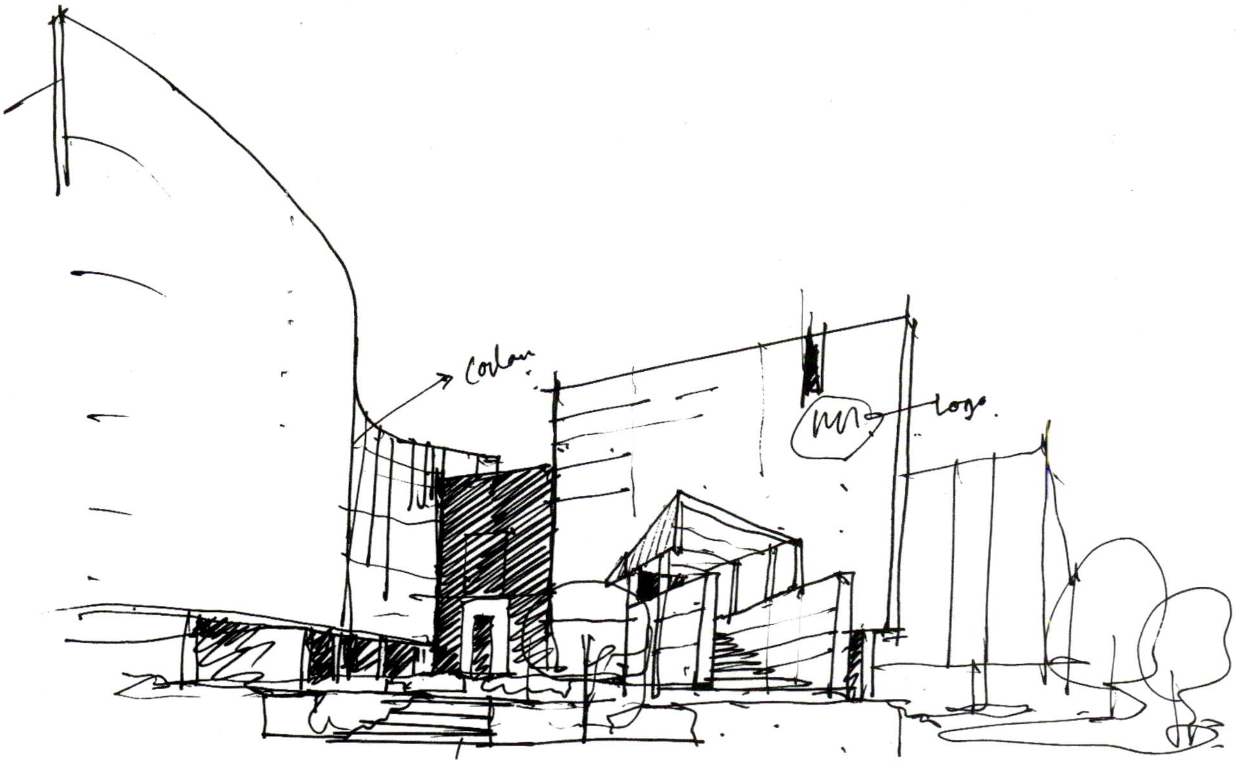

Nature in its various forms has always been a constant source of inspiration to human. The sky with its various states and ever changing colors is one of the most dynamic elements of nature and is what has inspired the façade. The colored laminate with Vanceva PVB glass on the façade follows no repetitive pattern and is symbolic of the unpredictable nature of the sky. It also represents the vibrancy of today's youth.

The chromatic glass façade comprises of 2 layers of 5mm clear glass each with an intermediate PVB layer 1.2mm thick which acts as an insulator, reducing the heat buildup inside. The façade becomes more dynamic by night when the inside lights are switched on and has a mesmerizing effect on the plaza below.

Circulation spaces and spill out zones have been created along the curved glass façade, allowing them to be naturally lit in the day and giving them a more interesting and dynamic feel. Emphasis has been given to interaction points not just for students but also for the faculty encouraging exchange of knowledge and ideas at all levels.

The reception has been designed as a triple height space naturally lit by a huge skylight above. The corridors and spill out spaces culminate around here at the upper floor levels. Care has been taken to ensure that the library is also well lit during the day and has been designed such that it is surrounded by glass on three sides, shaded by the existing trees on site. Color has subtly been introduced in the interiors as well in the form of colored glass partitions, colored glass slits on the wall or simply in the furniture.

An integrated plaza with a water body has been designed. The water body receives no direct sunlight and creates a comfortable micro-climate. Wherever possible shading trees have been planted and soft landscape has been introduced.

It is an institute that has achieved international standards not only by providing state of art facilities but also by having an architectural expression which represents that.

ARCHITECT
Abin Chaudhuri (Principal), Jui Mallik, Kaushik Majumder

FIRM
Abin Design Studio

AREA
27000 m^2

PHOTOGRAPHER
Pradip Sen

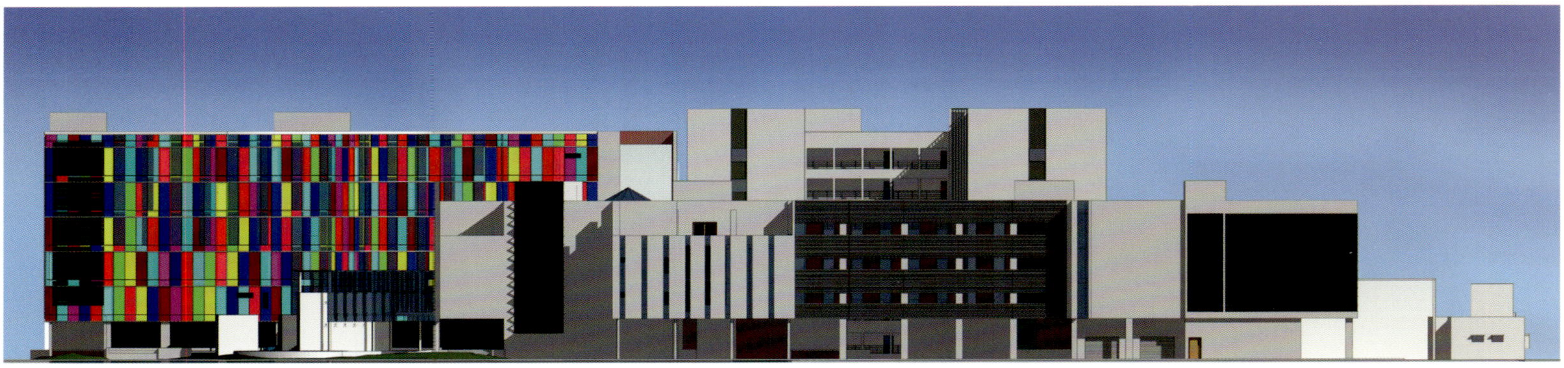

west elevation

north elevation

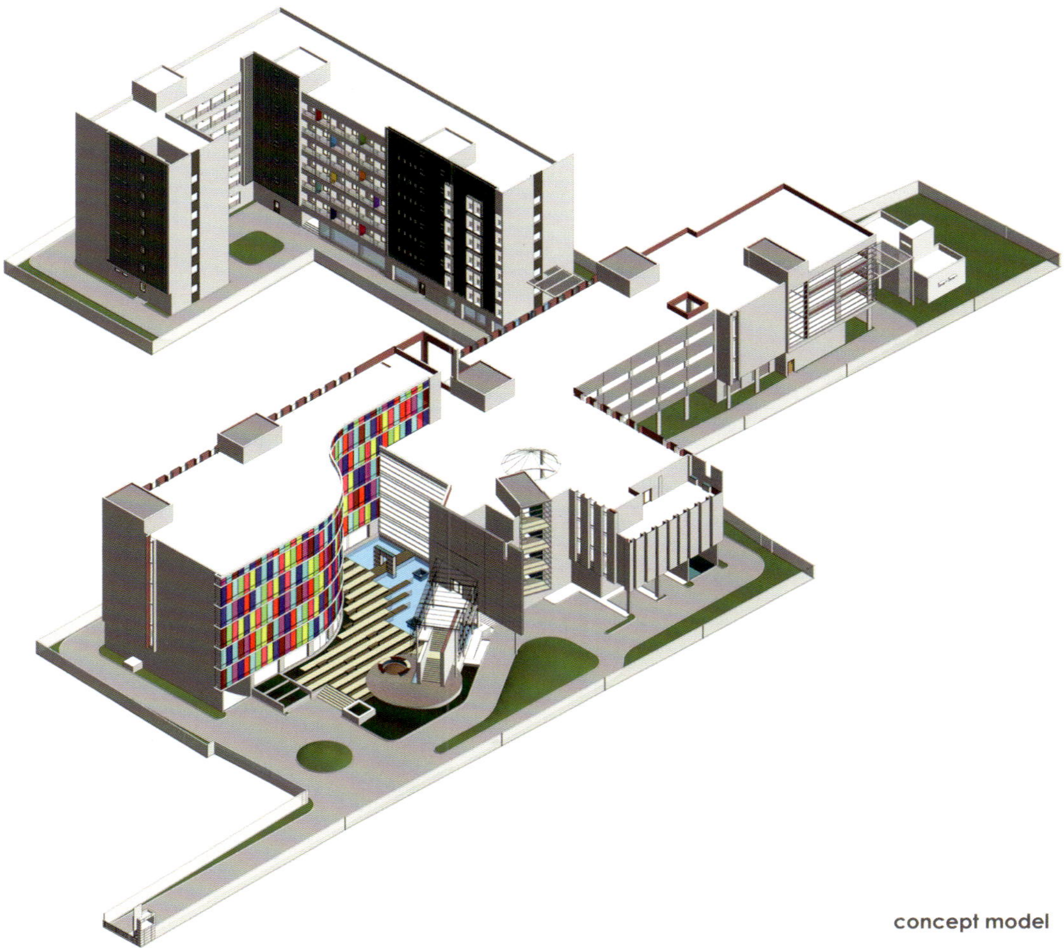

concept model

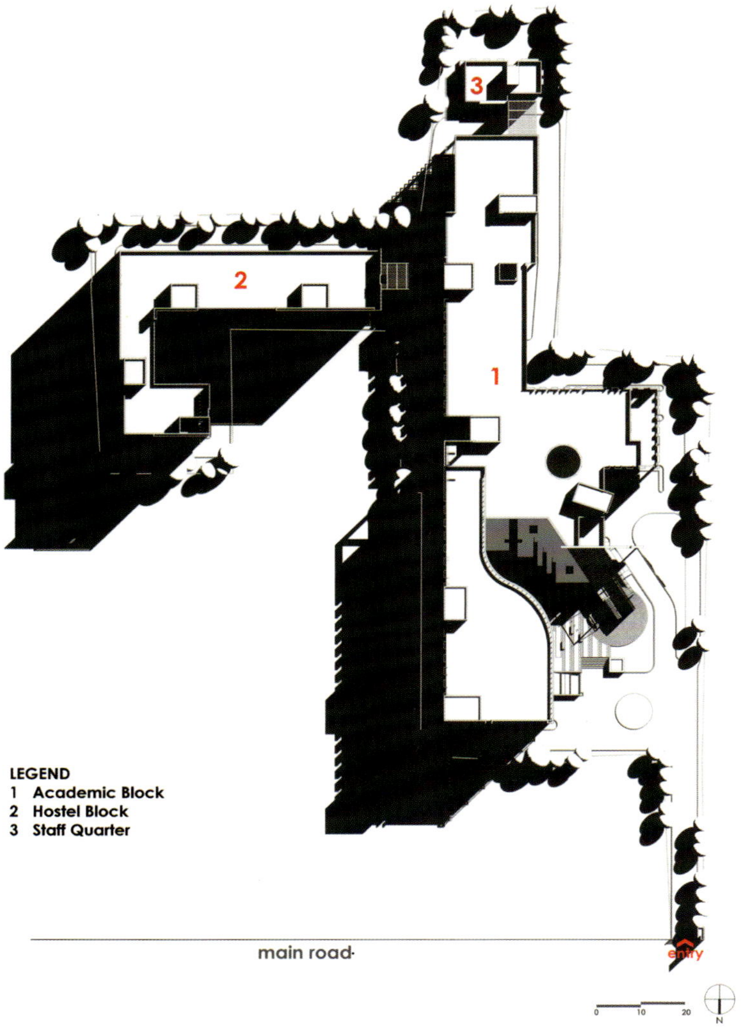

site plan

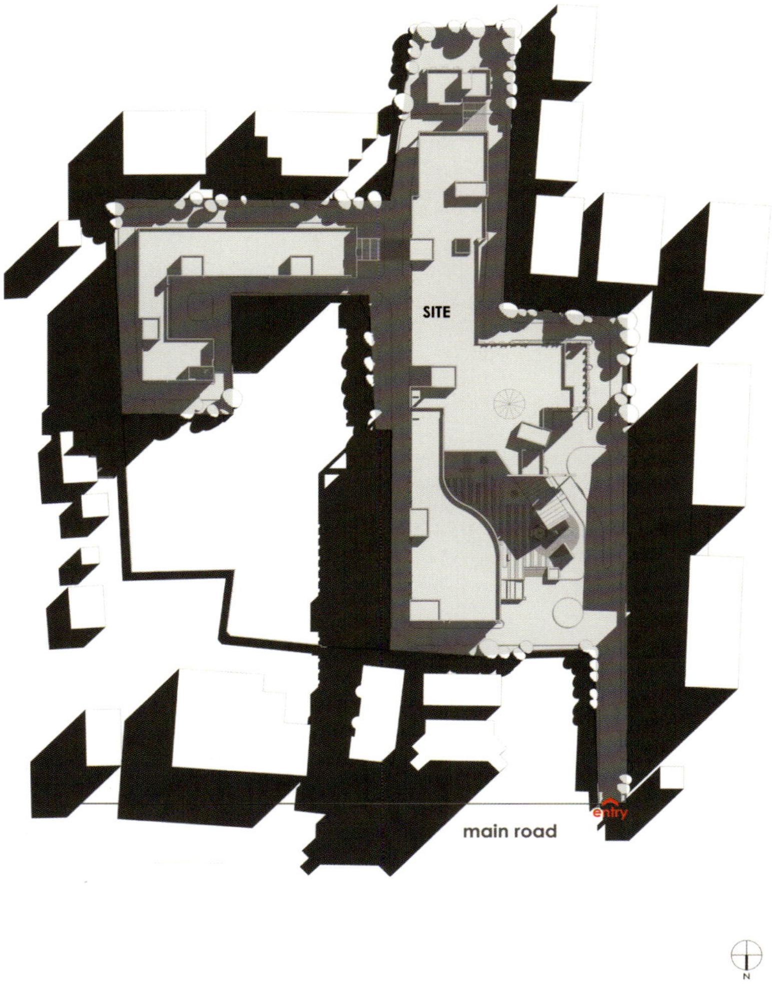

site & surroundings

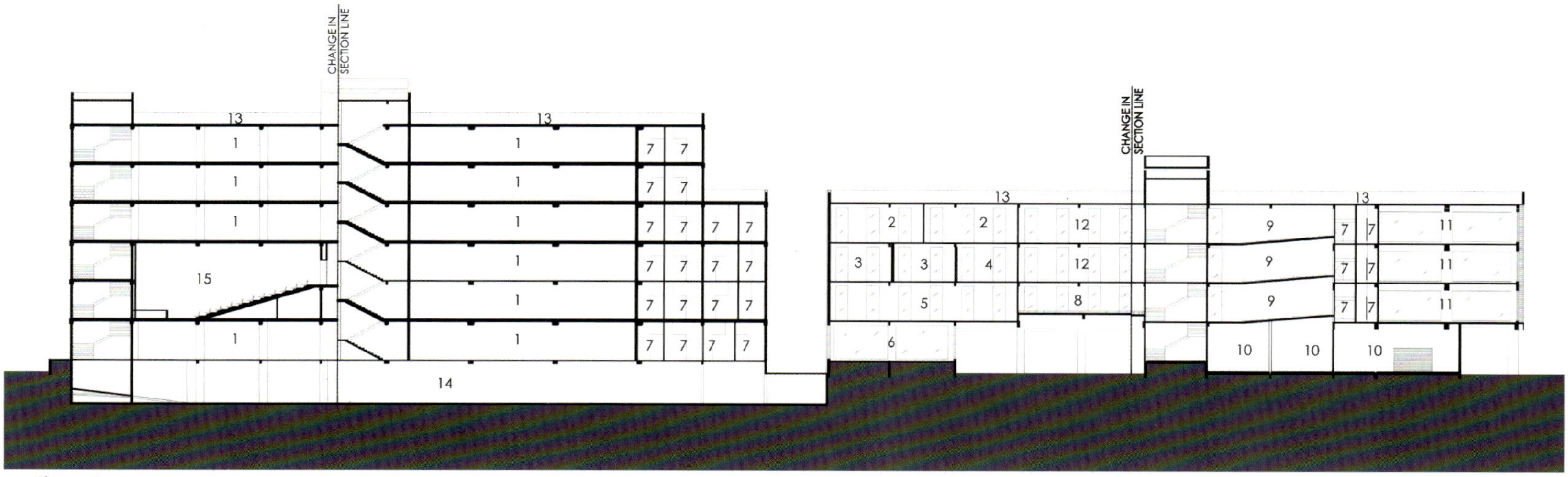

section A-A

LEGEND:
1 Office
2 Lecture Hall (Exec. Program)
3 Syndicate Room
4 Medical Room
5 Computer Room
6 Gymnasium
7 Toilets
8 Large Lecture Hall
9 Lecture Hall
10 Services
11 Library
12 Multi Purpose Hall
13 Open Terrace
14 Car parking
15 Convention Hall

自然总是以它不断变幻的形式成为人们设计的灵感来源。天空就是一个例子，它不断变化的颜色和状态就是这个设计立面的灵感来源。彩色薄片加上 PVB 玻璃使建筑立面生动活泼，上面的图案也是非重复的，象征着捉摸不定的天空，也表达着年轻人的无限活力。

玻璃立面色彩鲜艳，并以两层 5mm 的玻璃组成，每一层都有 1.2mm 的绝缘 PVB 涂层，以减少阳光给内部空间带来的热量。到了夜晚屋内点灯之后，立面显得更加光彩夺目，在下面的广场上渲染出迷人的色彩。

沿着弧形玻璃立面建有流通和溢出区域，白天采光充足，视觉上也动感十足，富有生趣。整个建筑重视公共交流场所的设计，不仅学生之间可以充分互动，还可随时与教员进行知识和观念的交流。

接待区高三层，以巨型天窗引入自然光照。图书馆的三边都是玻璃，馆内白天自然光线充足，旁边种植的大树为其遮挡阳光。除了立面、内部空间也以不同颜色的玻璃来分区，甚至家具也用色彩丰富的玻璃来做装饰。

设计还包括一个综合广场。广场上有避免了太阳直射的水景，周围设计有园林景观，种植了可以遮挡阳光的树木，营造出温馨舒适的环境。

整个学院的设计不仅使用了先进的材料和设备，而且很好地用建筑表达了符合国际标准的理念和形象。

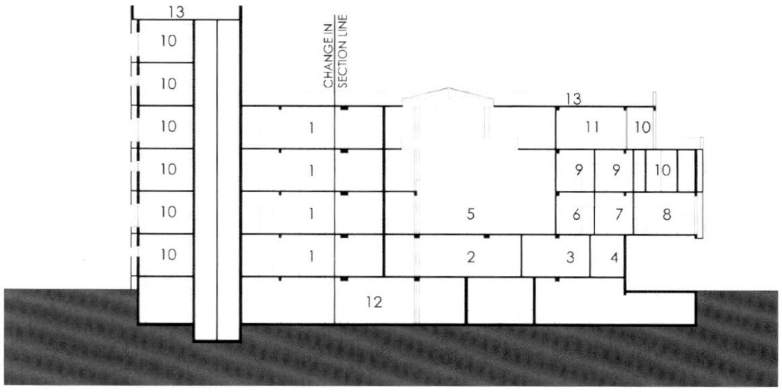

section B-B

LEGEND:
1 Office Lobby
2 Cafeteria
3 Kitchen
4 Store
5 Reception & Waiting Area
6 Registrar's Cabin
7 Dean's Cabin
8 Director's Cabin
9 Faculty Cabin
10 Toilet
11 Chairman's Office
12 Car Parking
13 Open Terrace

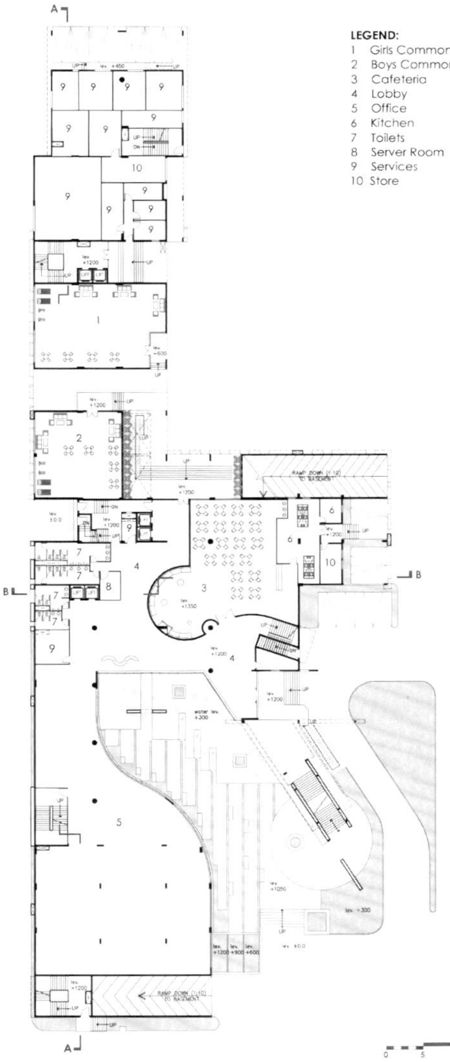

LEGEND:
1 Girls Common Room
2 Boys Common Room
3 Cafeteria
4 Lobby
5 Office
6 Kitchen
7 Toilets
8 Server Room
9 Services
10 Store

ground floor plan

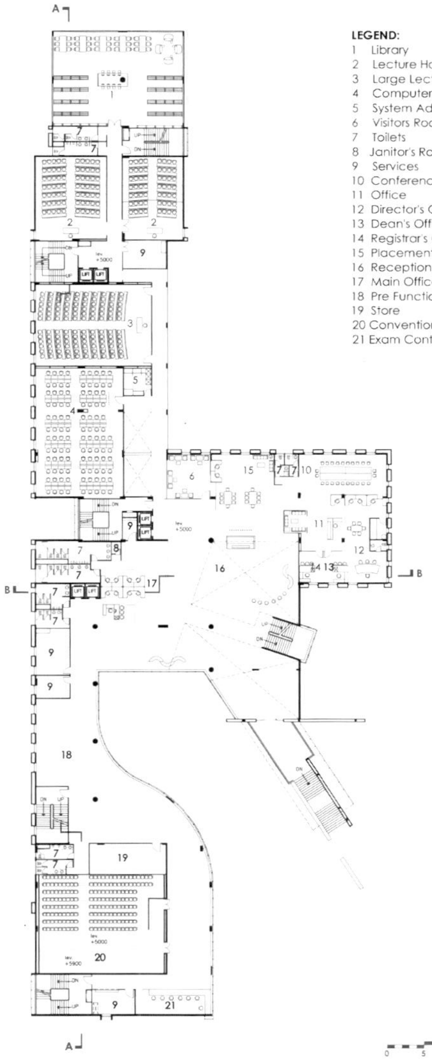

LEGEND:
1 Library
2 Lecture Hall
3 Large Lecture Hall
4 Computer Room
5 System Admin Room
6 Visitors Room
7 Toilets
8 Janitor's Room
9 Services
10 Conference Room
11 Office
12 Director's Office
13 Dean's Office
14 Registrar's Office
15 Placement Office
16 Reception & Waiting
17 Main Office
18 Pre Function Area
19 Store
20 Convention Hall
21 Exam Control Office

first floor plan

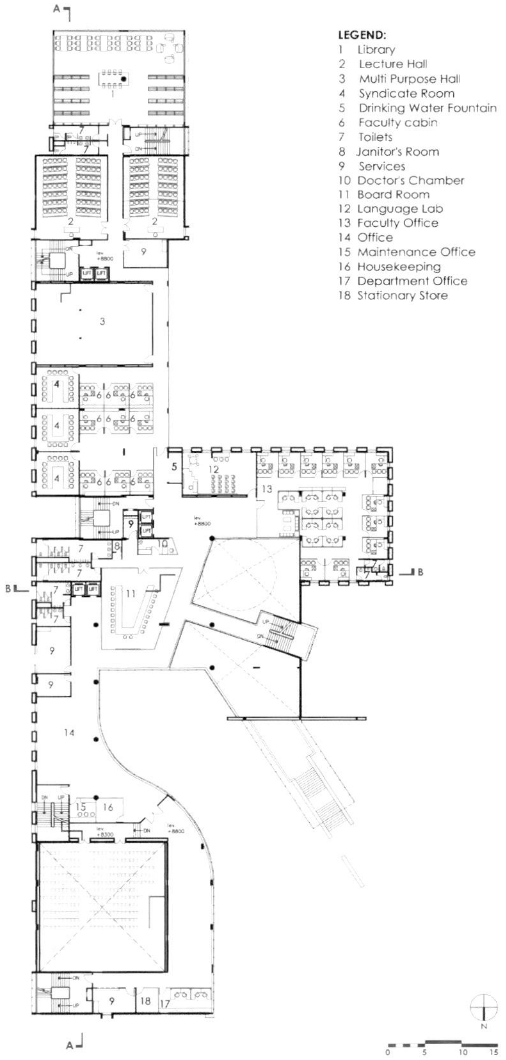

LEGEND:
1 Library
2 Lecture Hall
3 Multi Purpose Hall
4 Syndicate Room
5 Drinking Water Fountain
6 Faculty cabin
7 Toilets
8 Janitor's Room
9 Services
10 Doctor's Chamber
11 Board Room
12 Language Lab
13 Faculty Office
14 Office
15 Maintenance Office
16 Housekeeping
17 Department Office
18 Stationary Store

second floor plan

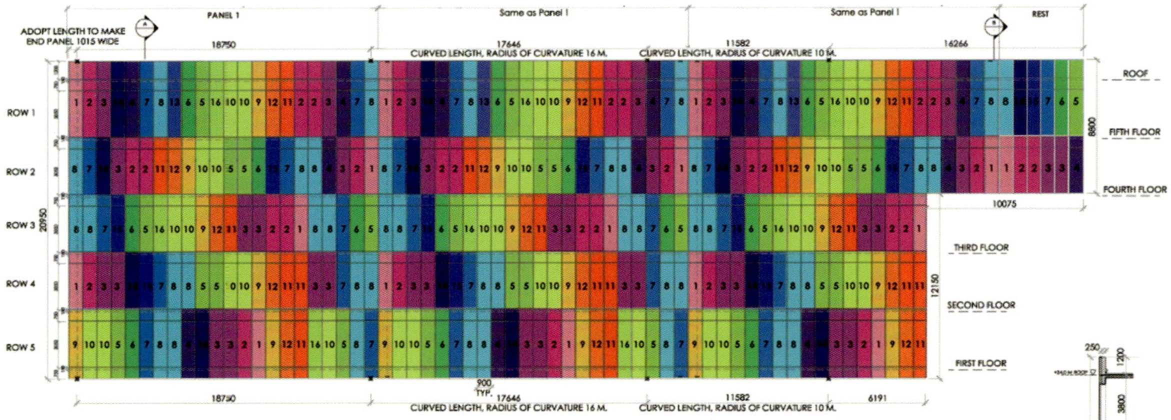

unfolded facade elevation

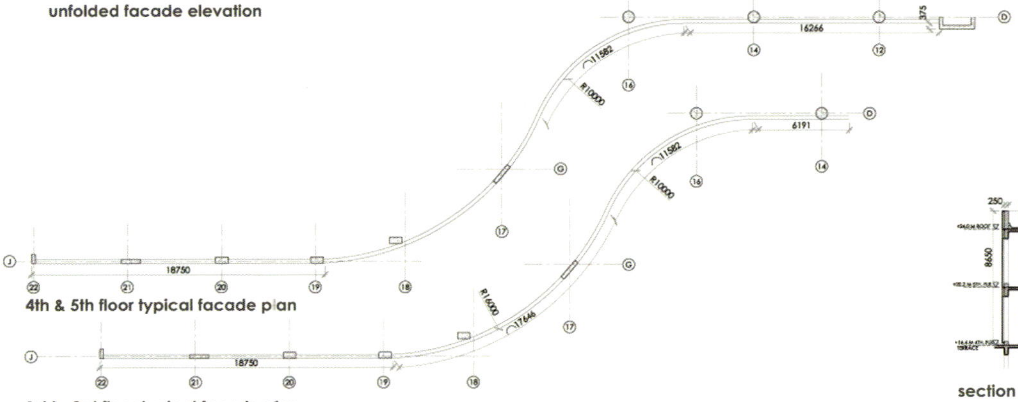

4th & 5th floor typical facade plan

1st to 3rd floor typical facade plan

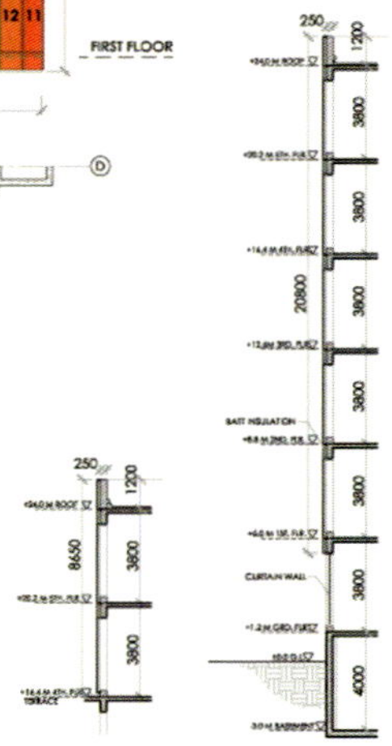

section B section A

color glass mapping

Homes Headquarters

The project of Mario Mazzer is characterized by two non-perpendicular intersected parallelepiped in order to express better space dialectics. The plan allocation bears in mind the concept of five elements and uses the water element faculty to balance the forces. In fact, the visitor is accompanied by a veil of water that runs right up to the entrance of the showroom which is positioned 6 meters below ground level. The main sides of the building are east and west facing exploiting the natural cycle of the sun. The main entrance is south facing.

The cross sectional building, overlaps the longitudinal structure that descends on the one side 6 meters under ground level, and offers the continuity of the façade by means of the curtain wall transparency that lines it. The double skin curtain wall façade, in extra-light glass is of the semi-active type, allowing high energy saving and is regulated by air extractors and curtains that offer excellent living wellbeing.

In the entrance there are two panoramic lifts and a steel and glass staircase characterized by finely shaped sides.

The last floor has a fantastic panoramic view of the surrounding hills, the view of the different floors and of the waterfall.

In the two underground floors of the building, there are garages, showroom, heating plant and warehouse. The outdoor floors always have transverse building reserved to reception and restrooms whereas in the longitudinal building there is the nursery, canteen with kitchen, offices and conference hall.

ARCHITECT
Mario Mazzer

TEAM PROJECT
Arch. Marco Da Ros- arch. Stefania Neodo

TEAM COMPETITION
Architects Marco Mirisola - Andrea Sartori - Luca Donati
Francesca Castagna - Alessandro Vergot - Gianluca Moras

STRUCTURE
Eng. Gianluca Sartori

CONSTRUCTION COMPANY
Tonon spa

AREA
14000 m²

LOCATION\Pieve di Soligo, Treviso, Italy

The layout of the offices with open space divisions, is flexible and easily adaptable to future needs thanks to floating floors and partition walls. The ceilings in micro perforated sheets promote high acoustic comfort.

An generous edge of land on the road front, produces sort of an amphitheatre that descends to the showroom and permits the natural light to illuminate it. Two deep shafts bring light centrally to the offices. A 9-meter drop of the first floor, determines the covered area assigned to the crèche and the canteen.

The lighting project collaborated technically to highlight the singularity of the architecture and supplies the work environment with scrupulous quality.

Much attention was in fact given to combining natural light with artificial light. The artificial light inside the building adjusts automatically, operating on time of day and on weather conditions.

The indoor areas are characterized by light colors. The showroom should make the core of the group understood, and through a high and dark shell highlight the quality and capture the imagination of visitors of the high development project towards which the five companies are launched.

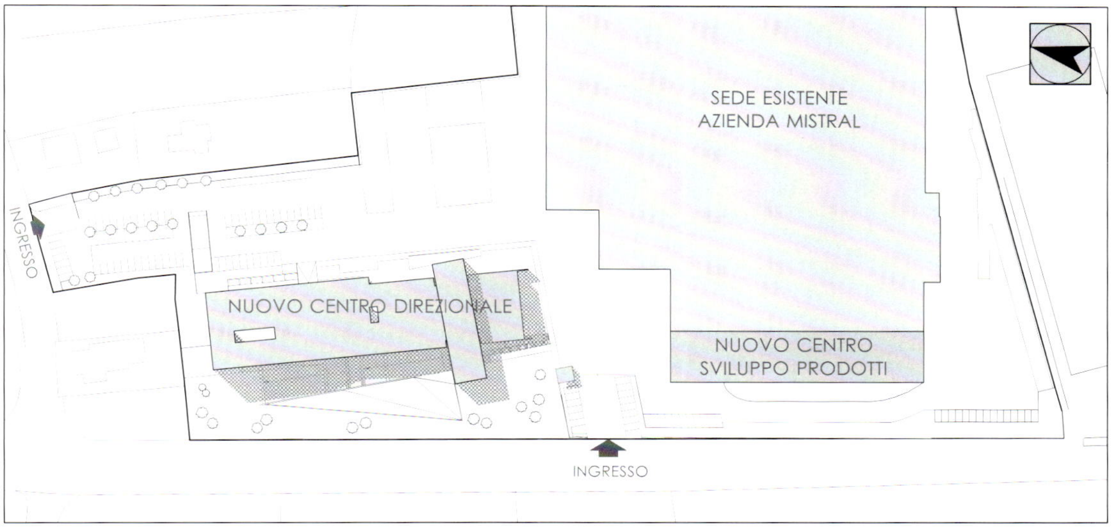

- PLANIMETRIA GENERALE - SCALA 1:2000
SITE PLAN - SCALE 1:2000

CONCEPT

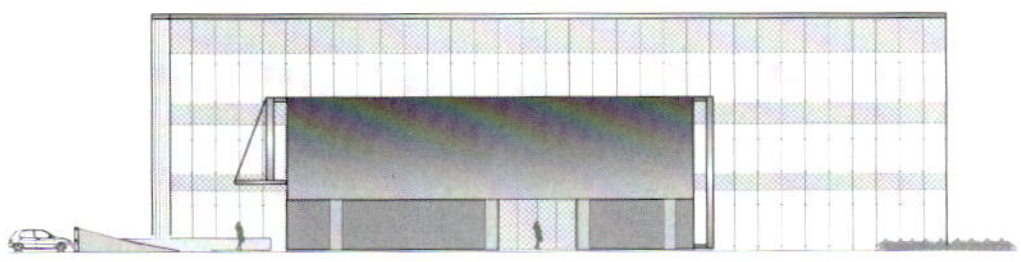

▪ PROSPETTO NORD - SCALA 1:500
NORTH ELEVATION - SCALE 1:500

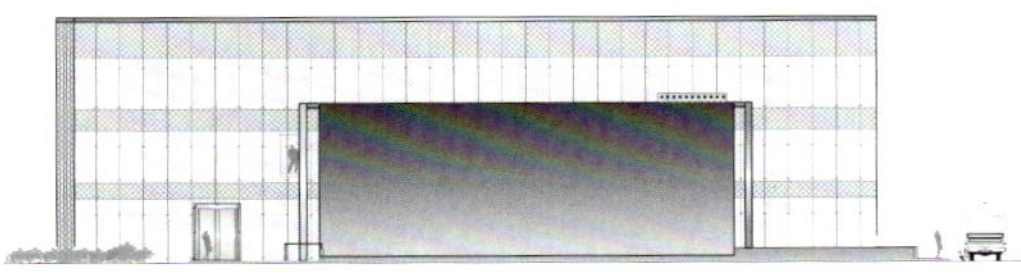

▪ PROSPETTO SUD - SCALA 1:500
SOUTH ELEVATION - SCALE 1:500

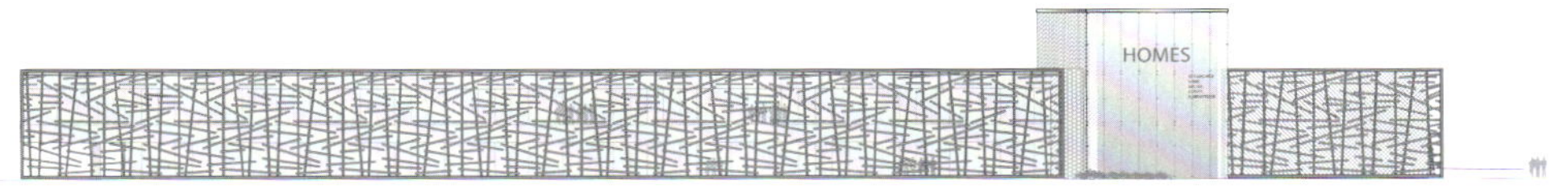

▪ PROSPETTO OVEST - SCALA 1:500
WEST ELEVATION - SCALE 1:500

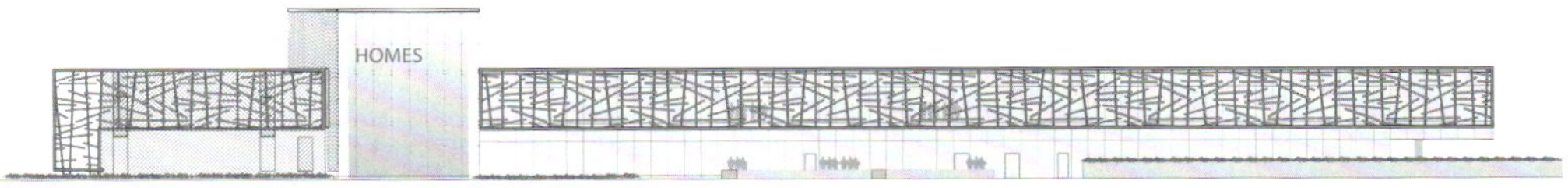

▪ PROSPETTO EST - SCALA 1:500
EAST ELEVATION - SCALE 1:500

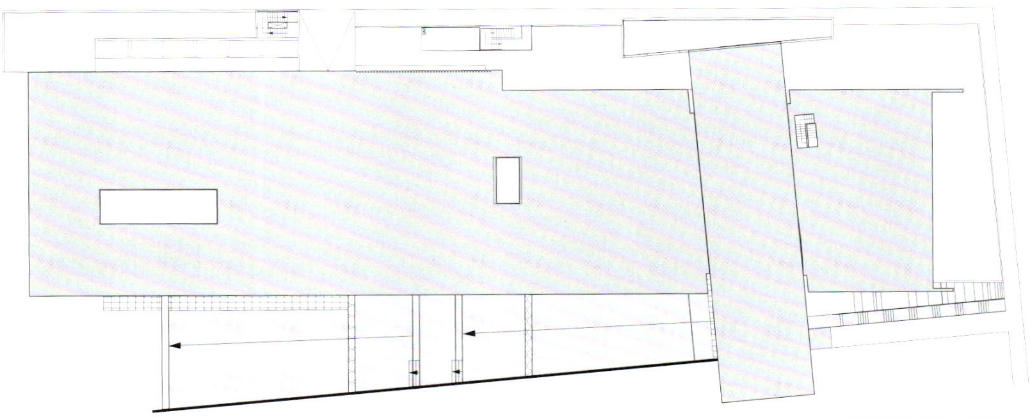

▪ PLANIMETRIA - SCALA 1:500
SITE PLAN - SCALE 1:500

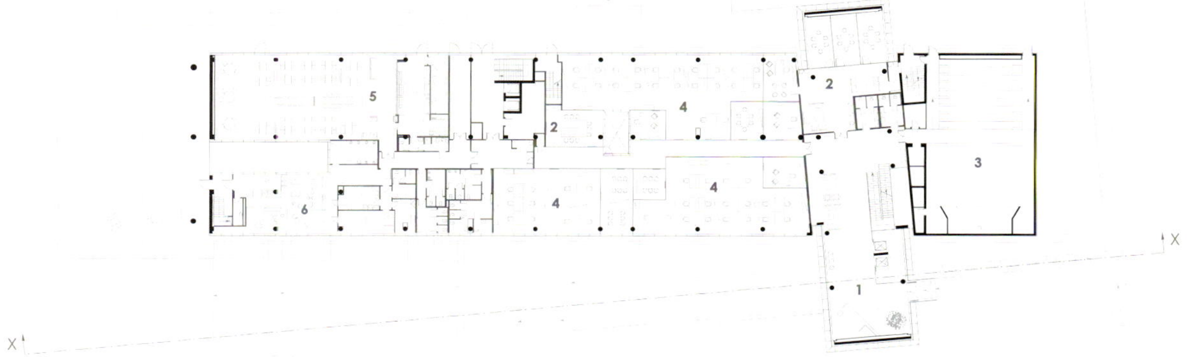

1. HALL/RECEPTION
2. SALE MEETING
3. SALA CONFERENZE
4. UFFICI
5. MENSA
6. NIDO AZIENDALE

1. HALL/RECEPTION
2. MEETING ROOM
3. CONFERENCE ROOM
4. OFFICES
5. RESTAURANT
6. NURSERY

▪ PIANTA PIANO TERRA - SCALA 1:500
GROUND FLOOR PLAN - SCALE 1:500

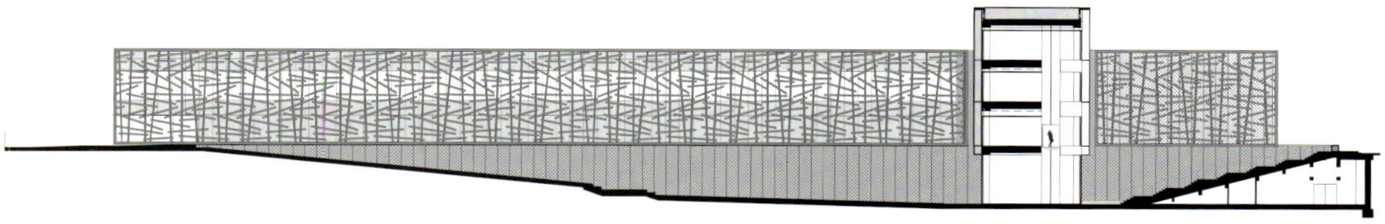

▪ SEZIONE LONGITUDINALE XX - SCALA 1:500
XX LONGITUDINAL SECTION - SCALE 1:500

建筑师 Mario Mazzer 设计的这个项目的主要特色是：2 座长方体的建筑成非直角状态相交，表现了一种空间辩证逻辑。该项目在规划布局时充分考虑了中国传统五行理论。在整个设计中用五行中的“水”来平衡整座建筑。事实上，来访者一走进该建筑群就一直有水相伴，水以一种形态一直流动到陈列室，陈列室的入口在地面下 6 米的位置。建筑物的主立面是东西向，充分考虑了太阳光的自然循环对建筑的影响。该建筑的大门向南。

两座相交的建筑中，横向大楼与纵向大楼重叠部分的楼体依地势一直向下延伸到地下 6 米处，透明的玻璃幕墙也一直随着建筑物向下。外立面由双层超轻玻璃制成，具有半主动控制功能，能很好地节约能源。地下室内的环境主要由抽气机进行调节，并装有内帘，使楼内具有良好的居住环境。在入口处有两部观光电梯和一个扶手楼梯，其造型优美，由钢铁与玻璃制成。

在建筑的最高层可以全方位地观赏到周围的山景，也可看到不同楼层的楼内景况和瀑布的景色。

在建筑的地下二层设有车库、陈列室、供暖设备和库房。横向建筑的一层有接待室和卫生间，纵向建筑里有育儿室，还有带厨房的餐厅、办公室和会议室。

办公区域的设置异常灵活，开放式的格子间里装有地板和隔墙，可根据用户的办公需要进行改装。天花板上安装的是微孔金属板，提高了听觉舒适度。

建筑外有很大一片空地形成巨大的户外广场，广场也随着地势一直向下延伸到陈列室前，陈列室就有了充足的自然光而明亮起来。两个电梯井的设置也使自然光线能进入到中央的办公区域。一层 9 米的落差也保证了足够的空间用于育儿室和餐厅。

结合了先进的照明技术，该设计更加突出了建筑的独特性，也使大楼内环境的每一个细节都显得一丝不苟。照明工程高度重视了自然光源与人工光源的配合，楼内的人工光源根据不同的时间段和天气条件能进行自动调节。

楼内以浅色调装饰为特色。核心区域是陈列室，其高耸的深色房屋架构突出了产品的质量，也很容易使来访者将其想象成高度发展的工程，目前已经有五家公司进驻于此。

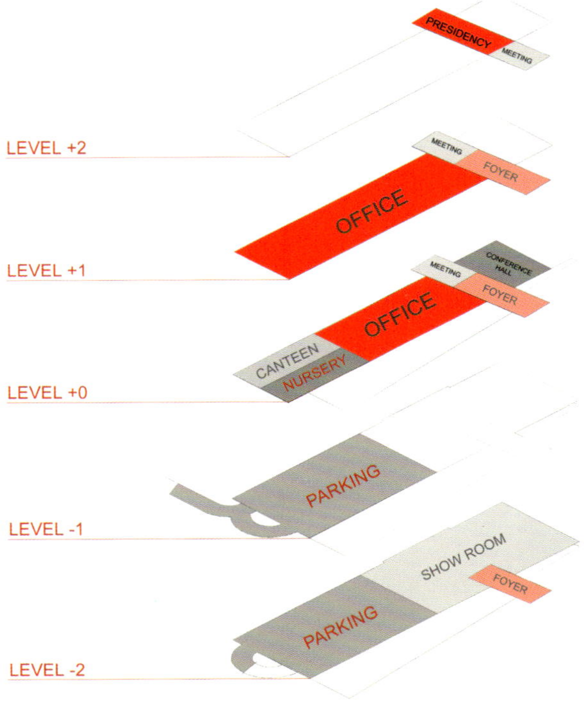

LAYOUT FUNCTIONAL

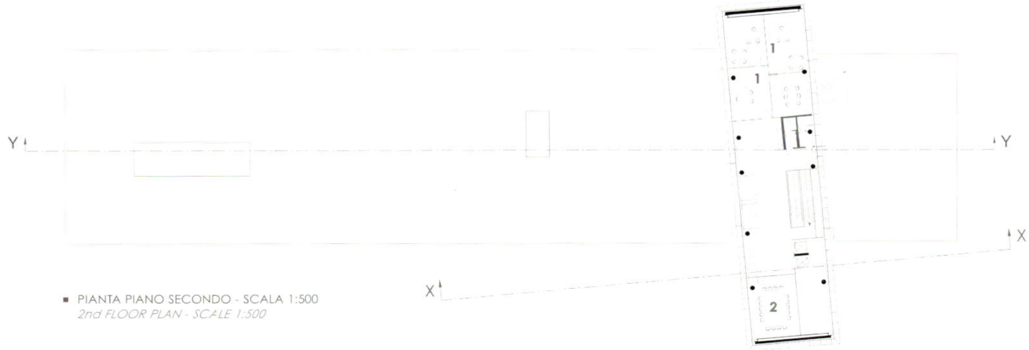

1. UFFICI
2. SALE MEETING

1. OFFICES
2. MEETING ROOM

■ PIANTA PIANO SECONDO - SCALA 1:500
2nd FLOOR PLAN - SCALE 1:500

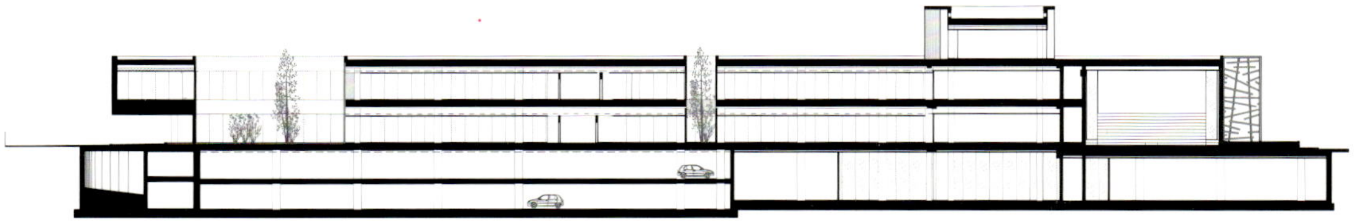

■ SEZIONE LONGITUDINALE YY - SCALA 1:500
YY LONGITUDINAL SECTION - SCALE 1:500

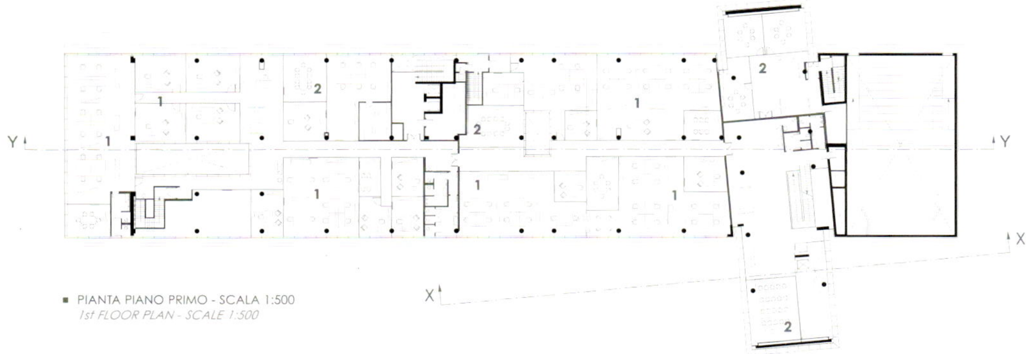

1. UFFICI
2. SALE MEETING

1. OFFICES
2. MEETING ROOM

■ PIANTA PIANO PRIMO - SCALA 1:500
1st FLOOR PLAN - SCALE 1:500

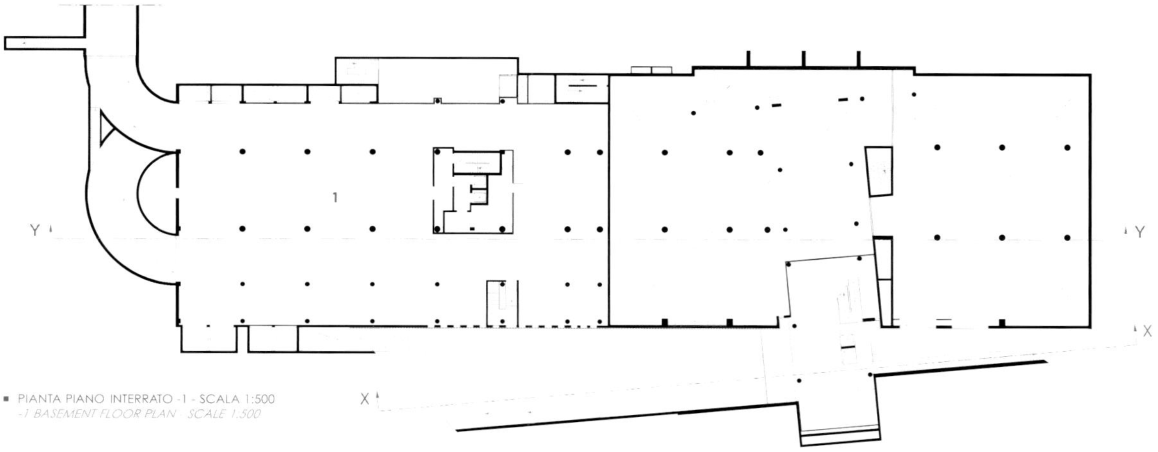

1. PARCHEGGIO

1. PARKING

■ PIANTA PIANO INTERRATO -1 - SCALA 1:500
-1 BASEMENT FLOOR PLAN - SCALE 1:500

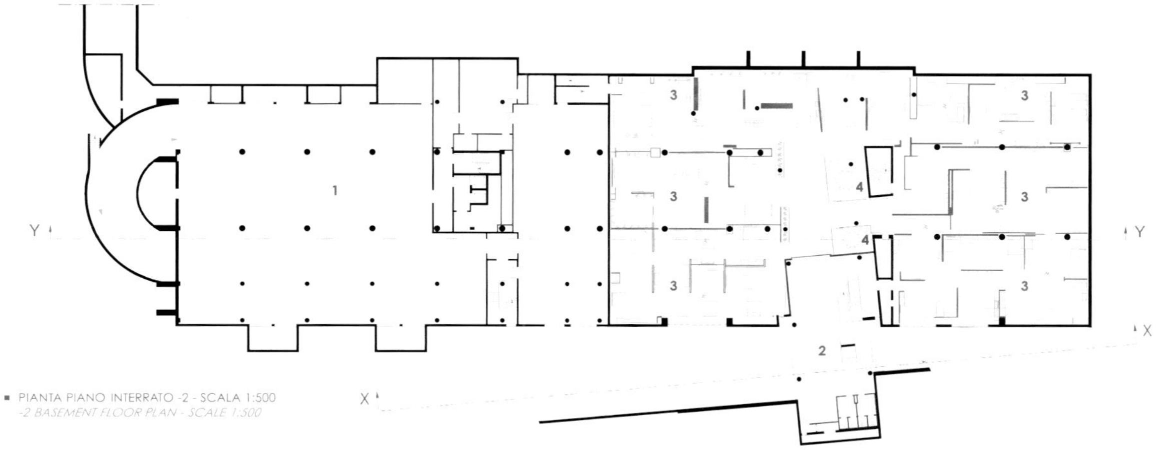

1. PARCHEGGIO
2. HALL/RECEPTION
3. SHOWROOM DELLE 5 AZIENDE DEL GRUPPO
4. SALE MEETING

1. PARKING
2. HALL/RECEPTION
3. SHOWROOMS
4. MEETING ROOM

■ PIANTA PIANO INTERRATO -2 - SCALA 1:500
-2 BASEMENT FLOOR PLAN - SCALE 1:500

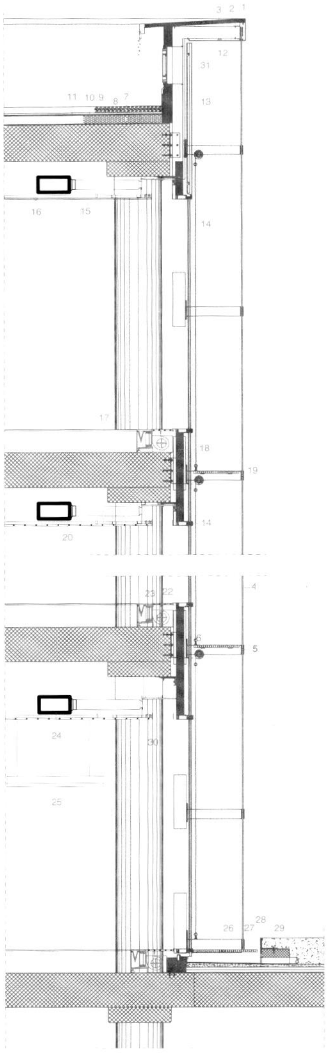

PARTICOLARE SEZIONE DOPPIA PELLE

scala 1:50

DETAIL: VERTICAL SECTION - DOUBLE SKIN

scale 1:50

1- scossalina in acciaio preverniciato
prepainted steel flashing
2- pannello impermeabilizzante
panel waterproofing
3- materiale isolante
insulating material
4- vetro extra chiaro sp.12 mm
extra clear glass 12 mm thick
5- profilo in accio inox AISI 304
profile in AISI 304 stainless steel
6- tenda esterna motorizzata
motorized awning
7- ghiaino
gravel
8- plateau proteggi guaina
peep toe protector sheath
9- doppia guaina bituminosa
double sheath of bitumen
10- isolamento sp.120 mm
insulation 120 mm thick
11- foglio di politilene
polythene sheet
12- led line philips RGB integrati sul pannello
philips integrated RGB LED-line panel
13- lamiera in acciaio preverniciato microforata
prepainted steel sheet perforated
14- vetro selettivo PLATIN 71/40 8 temperato/16/4+4 pvb 0.76 apribile ad anta verso l'interno
selective glass PLATIN 71/40 8 +4 temperato/16/4 pvb 0.76 openable door inwards
15- controsoffitto in topacustic, impiallacciato con essenza legnosa
topacustic ceiling, wood with wood veneer
16- illuminazione a led philips
LED lighting philips
17- pavimento in legno impiallacciato
wood veneer flooring
18- vetro PLATIN 71/40 sp.8 mm temperato 16 di camera e flot sp.5 mm acidato temperato camera ventilata 60, isolamento sp.70 mm
71/40 8 mm tempered glass PLATIN 16 bed and floating 5 mm thick tempered frosted ventilated chamber 60, isolation sp.70 mm
19- mensole passanti in acciaio zincato
through galvanized steel shelves
20- controsoffitto in cartongesso
plasterboard ceiling
21- placca di finitura in acciaio preverniciato, giunto di ancoraggio dell'infisso, infisso termoisolante in acciaio
plate coated steel finish, joint the anchor of the casing steel frame insulation
22- griglia in acciaio inox
stainless steel grill
23- ventil convettore da pavimento AERNOVA
floor fan coil AERNOVA
24- unità di trattamento aria AERNOVA
air handling units AERNOVA
25- apparecchi philips illuminazione dinamica
philips dynamic lighting equipment
26- led line philips RGB integrati su griglia
philips integrated RGB LED line on the grid
27- griglia acciaio inox
stainless steel grill
28- vasca raccolta acqua e ripresa aria
water collection tank and air intake
29- tubo smaltimento acqua piovana
rainwater disposal pipe
30- veletta in cartongesso con neon integrato
veil plasterboard with integrated neon
31- ventola espulsione aria calda
hot air expulsion fan

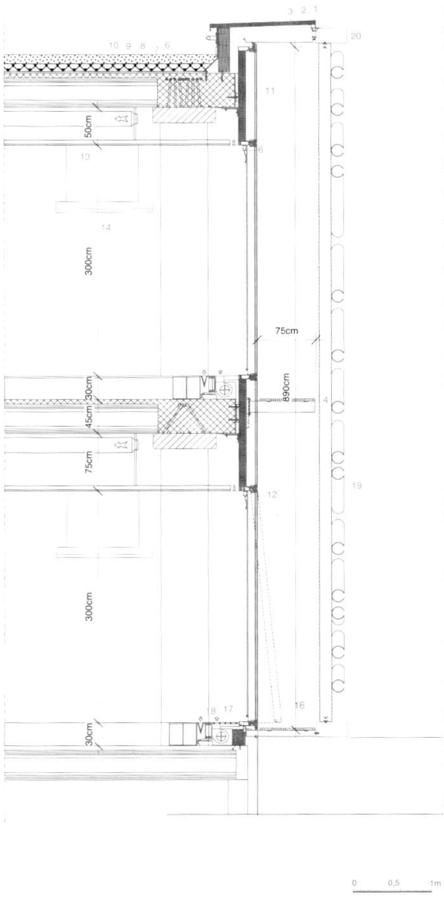

PARTICOLARE SEZIONE SU FACCIATA CORPO UFFICI

scala 1:50

DETAIL: VERTICAL SECTION - OFFICES FACADE

scale 1:50

1- scossalina in acciaio preverniciato
prepainted steel flashing
2- pannello impermeabilizzante
panel waterproofing
3- materiale isolante
insulating material
4- mensole passanti in acciaio zincato
through galvanized steel shelves
5- tenda interna motorizzata
motorized awning
6- ghiaino
gravel
7- plateau proteggi guaina
peep toe protector sheath
8- doppia guaina bituminosa
double sheath of bitumen
9- isolamento sp.120 mm
insulation 120 mm thick
10- foglio di politilene
polythene sheet
11- vetrocamera Sunbelt Arctis Green 43/20 temperato 8/16/5 temperato acidato, retrostante pannello in lamiera di acciaio verniciato con incollato materassino in lana minerale sp.80 mm e lamiera skinplate a finire
Green glazed Sunbelt Arctis 43/20 8/16/5 toughened tempered glass, the rear panel of sheet steel painted with mineral wool mat glued sp.80 mm and sheet skinplate end
12- vetrocamera Sunbelt Arctis Green 43/20 temperato 8/16/5 apribile ad anta verso l'esterno
Sunbelt Arctis Green tempered glazing 8/16/5 43/20 openable door to the outside
13- controsoffitto in lamiera microforata
perforated metal ceiling
14- apparecchi philips illuminazione dinamica
philips dynamic lighting equipment
15- pavimento sopraelevato in gres porcellanato
raised floor porcelain tile
16- griglia in accio inox
stainless steel grill
17- ventil convettore da pavimento AERNOVA
floor fan coil AERNOVA
18- unità di trattamento aria AERNOVA
air handling units AERNOVA
19- texture d'alluminio
texture aluminum
20- struttura texture d'alluminio
texture structure of aluminum

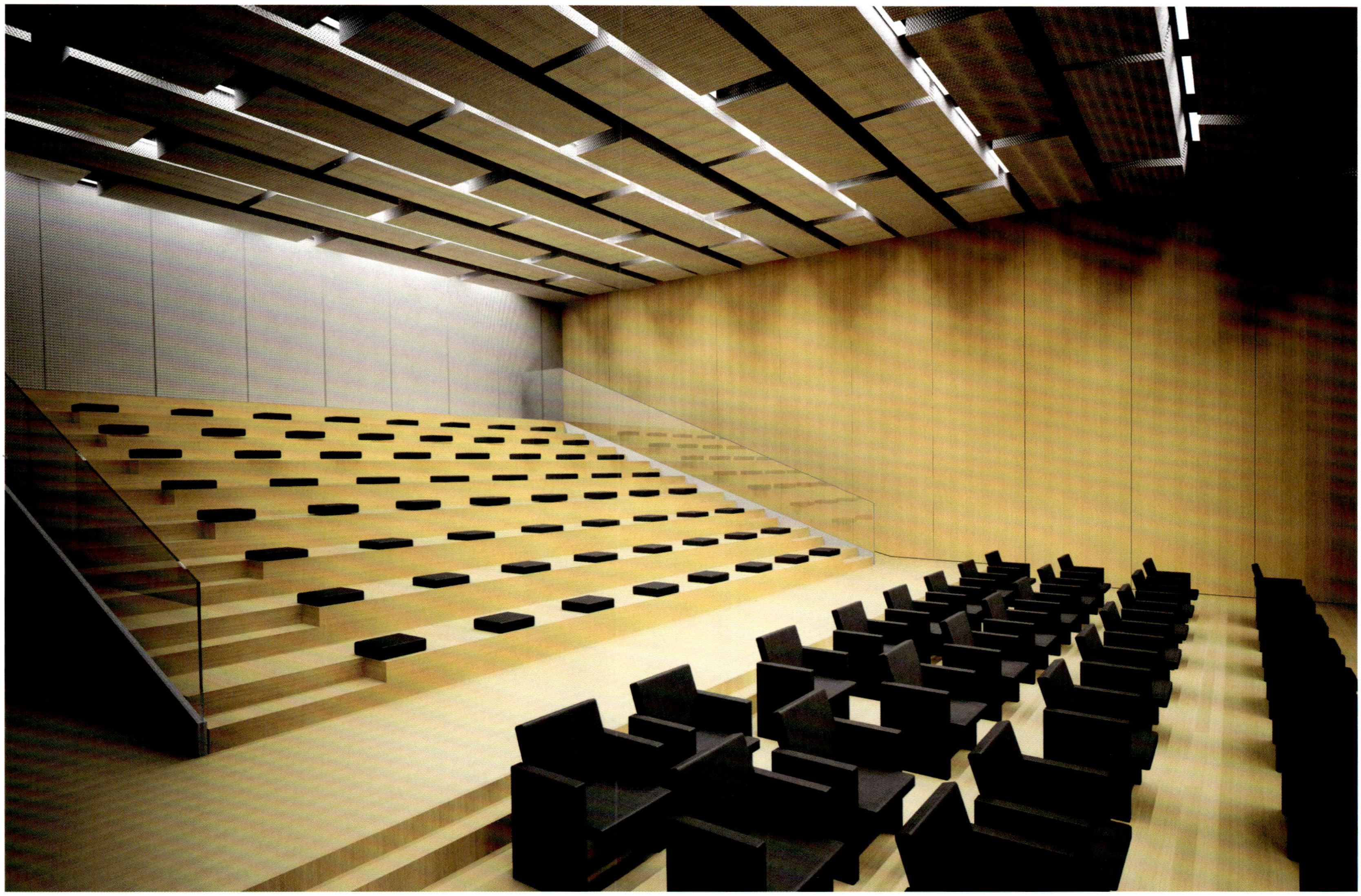

Man Air

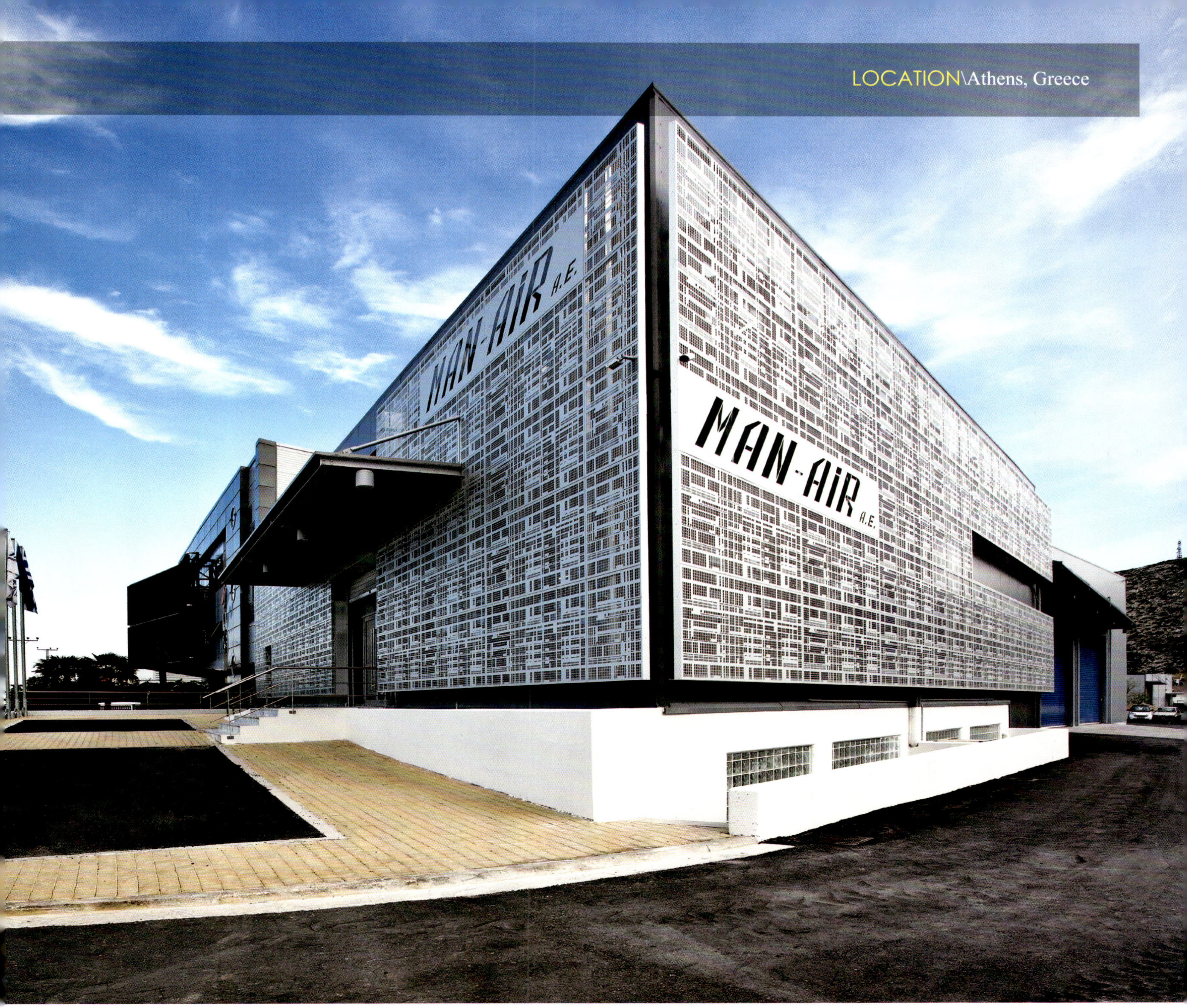

LOCATION\Athens, Greece

ARCHITECT
Sotiris Tsoulos
FIRM
Tsoulos Associates Ltd.
AREA
1100 m^2

Man Air S.A. is a Greek company that deals heavy machinery for construction ,drilling and earthmoving, the company has been looking for its new headquarter where it would combine in one space both the technical support and storage with the management. The company deals directly with construction firms.

The program asked for a complex integration of management offices and technical spaces in one structure. In total there was necessity of 22 offices with board room, meeting rooms, reception and 500 square meters of machinery room connected with a heavy material storage space a basement and an outdoor working space. The machinery has to be noted that would be up to 29 meters long.

The building was divided in two adjacent buildings with separate structure. The north is a 2 floors high building with basement and angular to fit the site and accommodates all the offices and common spaces. The south building is a rectangular box with strategically placed openings to allow the heavy equipment to enter and create 2 working positions or 1 that spans through the entire building.

The concept was to create a space that would fit the company profile, a healthy Greek company that deals and services heavy machinery. The idea was to create a building that looks like a well oiled machine and depicts and compliments the activities of the firm. The details were not hidden or claded, the floors were industrial and the ceilings of corrugated metal panels were painted as well as the beams. The machine of course needs a cover and the skin that was chosen was anodized aluminum panels, punctured to a pattern that noted the secondary digital identity of the company. Great attention was paid to all the visible details. The electricity channels were all designed beforehand and hanged with precision.

Lighting has been thoroughly examined and the perforated façade assists in steady shading with approximately 50% opacity. The interior lighting fixtures were linear and in the ground floor hang in different height and lengths to recreate the daylight penetration in an artificial way. The 3D effect that we get from the different heights of hanging the light fixtures assists the understanding of the volume. The central staircase was claded with the same punctured aluminum elements and light from within with LED lights plus from the side to accent the "object in space" effect. The floors are all with industrial self leveling resin lightly colored.

1. ECOPANEL WLC
2. ΕΙΔΙΚΟ ΤΕΜΑΧΙΟ ΣΤΗΡΙΞΗΣ
3. ΑΥΤΟΔΙΑΤΡΗΤΙΚΗ ΒΙΔΑ ΣΥΝΕΝΩΣΗΣ 5,5mm x L ΚΑΙ ΜΕΤΑΛΛΙΚΗ ΡΟΔΕΛΑ ΜΕ EPDM
4. ΕΙΔΙΚΟ ΤΕΜΑΧΙΟ
5. ΤΕΓΙΔΑ ΣΧΗΜΑΤΟΣ «C»
6. ΚΟΥΦΩΜΑ ΑΛΟΥΜΙΝΙΟΥ
7. ΜΕΤΑΛΛΙΚΗ ΚΟΛΩΝΑ ΣΧΗΜΑΤΟΣ" «H»
8. ΒΙΔΑ ΣΤΕΡΕΩΣΗΣ
9. ΕΙΔΙΚΟ ΤΕΜΑΧΙΟ ΚΟΥΦΩΜΑΤΟΣ
10. ΜΕΤΑΛΛΙΚΟ ΠΑΝΕΛΟ ΣΙΚΑΣΗΣ
11. ΣΚΕΛΕΤΟΣ ΠΑΝΕΛΩΝ ΣΚΙΑΣΗΣ
12. ΕΙΔΙΚΟ ΤΕΜΑΧΙΟ ΣΤΗΡΙΞΗΣ ΣΚΕΛΕΤΟΥ ΠΑΝΕΛΩΝ ΣΚΙΑΣΗΣ
13. ΑΥΤΟΔΙΑΤΡΗΤΙΚΗ ΒΙΔΑ ΣΤΕΡΕΩΣΗΣ 5,5-6,3mm x L ΚΑΙ ΜΕΤΑΛΛΙΚΗ ΡΟΔΕΛΑ ΜΕ EPDM
14. ΒΙΔΑ ΣΥΡΡΑΦΗΣ

ΤΟΜΗ

MAN-AIR A.E.

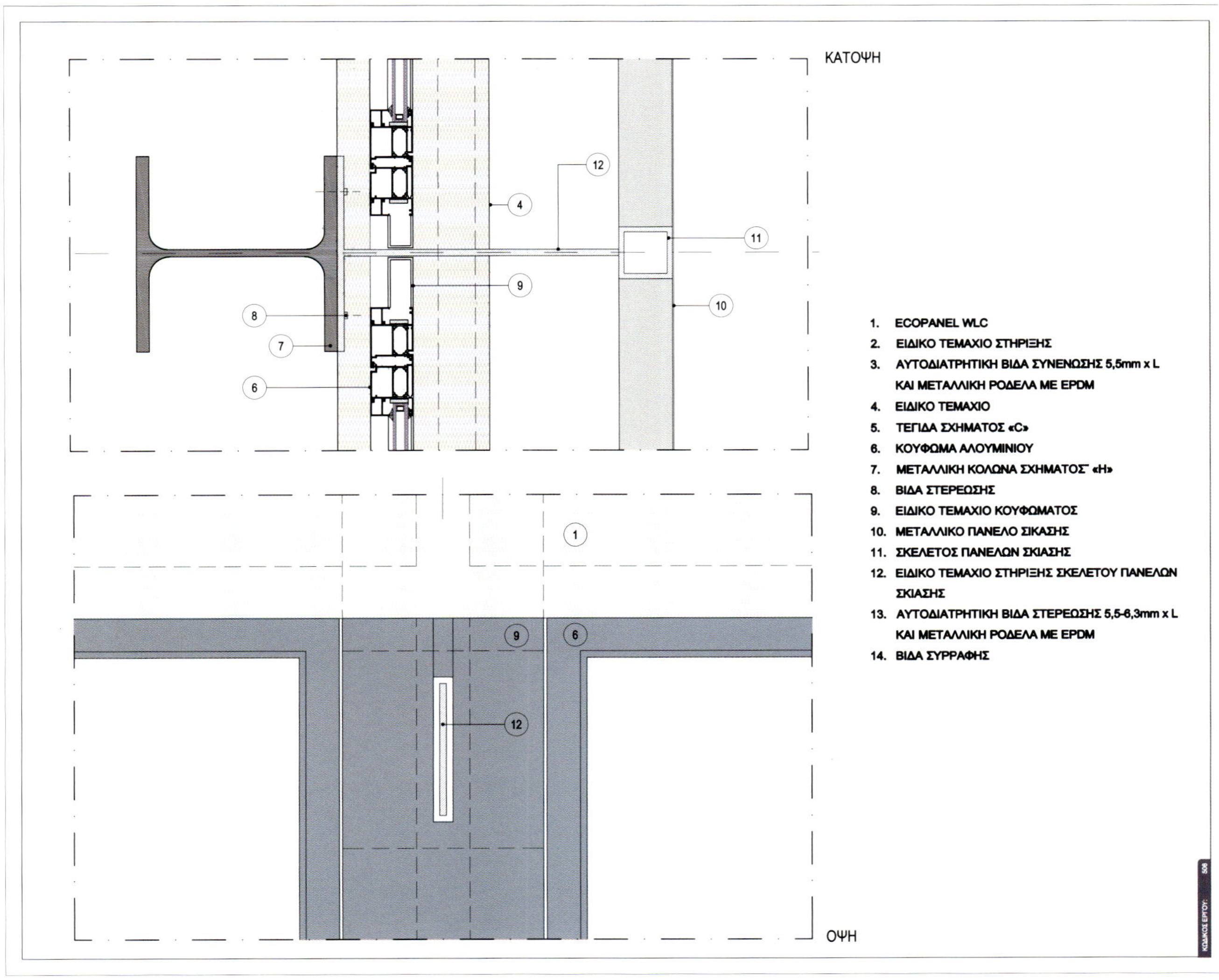
ΚΑΤΟΨΗ
1
4
6
7
8
9
10
11
12
1. ECOPANEL WLC
2. ΕΙΔΙΚΟ ΤΕΜΑΧΙΟ ΣΤΗΡΙΞΗΣ
3. ΑΥΤΟΔΙΑΤΡΗΤΙΚΗ ΒΙΔΑ ΣΥΝΕΝΩΣΗΣ 5,5mm x L ΚΑΙ ΜΕΤΑΛΛΙΚΗ ΡΟΔΕΛΑ ΜΕ EPDM
4. ΕΙΔΙΚΟ ΤΕΜΑΧΙΟ
5. ΤΕΓΙΔΑ ΣΧΗΜΑΤΟΣ «C»
6. ΚΟΥΦΩΜΑ ΑΛΟΥΜΙΝΙΟΥ
7. ΜΕΤΑΛΛΙΚΗ ΚΟΛΩΝΑ ΣΧΗΜΑΤΟΣ «Η»
8. ΒΙΔΑ ΣΤΕΡΕΩΣΗΣ
9. ΕΙΔΙΚΟ ΤΕΜΑΧΙΟ ΚΟΥΦΩΜΑΤΟΣ
10. ΜΕΤΑΛΛΙΚΟ ΠΑΝΕΛΟ ΣΙΚΑΣΗΣ
11. ΣΚΕΛΕΤΟΣ ΠΑΝΕΛΩΝ ΣΚΙΑΣΗΣ
12. ΕΙΔΙΚΟ ΤΕΜΑΧΙΟ ΣΤΗΡΙΞΗΣ ΣΚΕΛΕΤΟΥ ΠΑΝΕΛΩΝ ΣΚΙΑΣΗΣ
13. ΑΥΤΟΔΙΑΤΡΗΤΙΚΗ ΒΙΔΑ ΣΤΕΡΕΩΣΗΣ 5,5-6,3mm x L ΚΑΙ ΜΕΤΑΛΛΙΚΗ ΡΟΔΕΛΑ ΜΕ EPDM
14. ΒΙΔΑ ΣΥΡΡΑΦΗΣ
ΟΨΗ

MAN-AIR

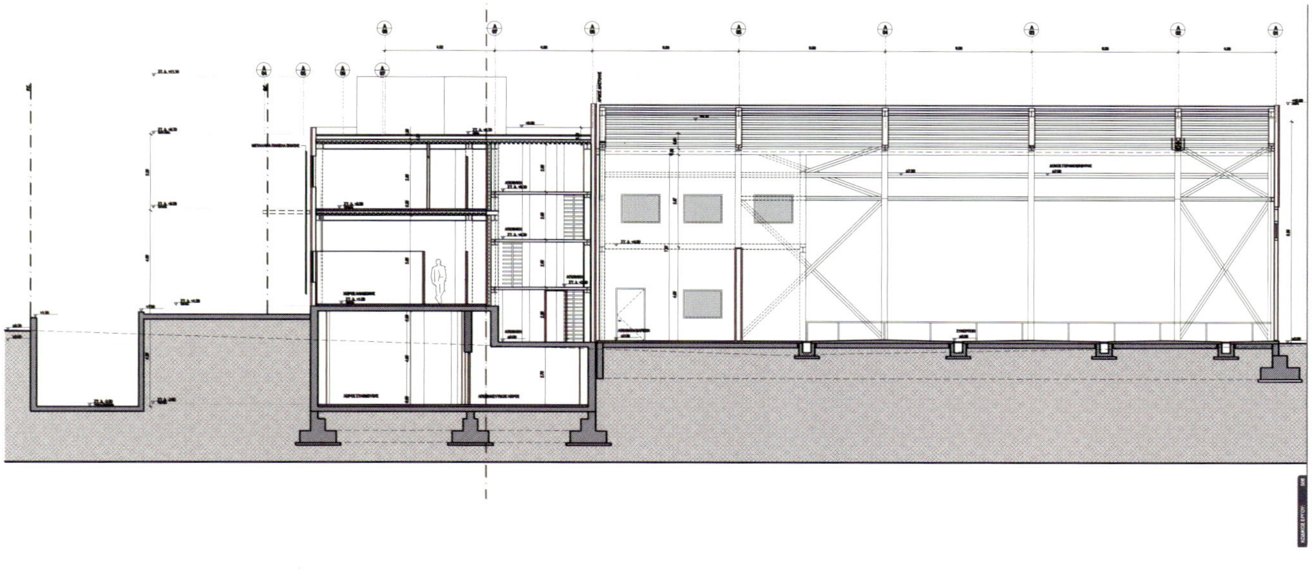

曼埃尔公司是希腊的一家公司，主要经营用于建筑、钻孔和土方工程的机械。该公司计划建造一个新的总部，要既能提供技术支持又能提供管理人员的办公空间。该公司直接与建筑公司联系。

基于公司的上述特点，其总部大楼就需要建成一座集管理人员办公场所和技术空间于一体的复合型综合建筑。工程总体上需要具有 22 个独立的办公场所用于董事会会议室、普通会议室和接待室，500 平方米的机房和与机房相通的重型机械储存空间，一个地下室和室外工作空间。要特别强调的是部分机械的长度会达到 29 米。

工程分为两个相邻的具有不同结构和用途的大楼，但相邻的两座大楼之间有两处连接口。北楼高为两层，设有地下室。为了适应地形，北楼呈现一定的角度，所有的办公区域和公共空间都坐落在北楼。南楼是一座方形的建筑。从设计策略上说，其四周都是开放的开阔空间以让重型设备进出，同时其内部开辟有两个工作空间或者一个横跨整座大楼的工作空间。

该工程的设计理念就是要创造一个符合公司形象的空间，曼埃尔公司是一家运转良好的希腊公司，其经营重型机械并提供售后服务。该工程就是要建造一座看起来像是运转良好的机器的大楼，用建筑本身来描述和称赞公司的经营和服务行为。建筑的细节部位既不掩藏也不覆盖，充分暴露在外。设计将各楼层工业化，波纹状金属面板制成的天花板和横梁都有喷漆。机器当然需要外壳和外立面，在工程设计中通过选择使用阳极氧化铝板外立面来实现，并将其切割成一定的图案以标记公司的第二大特征——数字化特性。设计对每一处可见的细节都给予了关注。电线管是提前设计好的，其悬挂的高度都有精密的计划；照明设备受到全面的检测，穿孔的外立面大约有 50% 的面积不透明，形成了稳定的遮阳设施。室内的照明设施呈线性分布，底层的照明设施长长短短，高高矮矮地悬挂着以创造出自然光线的效果。不同高度的照明设施营造出的立体效果增强了楼层空间的体量感。中央楼梯包有与立面同样的穿孔式阳极氧化铝板。来自内部和侧面的发光二极管的灯光增强了楼梯“悬在空中”的效果。铺地采用的是工业使用的浅色自流平树脂板。

55 Park Street Laboratory

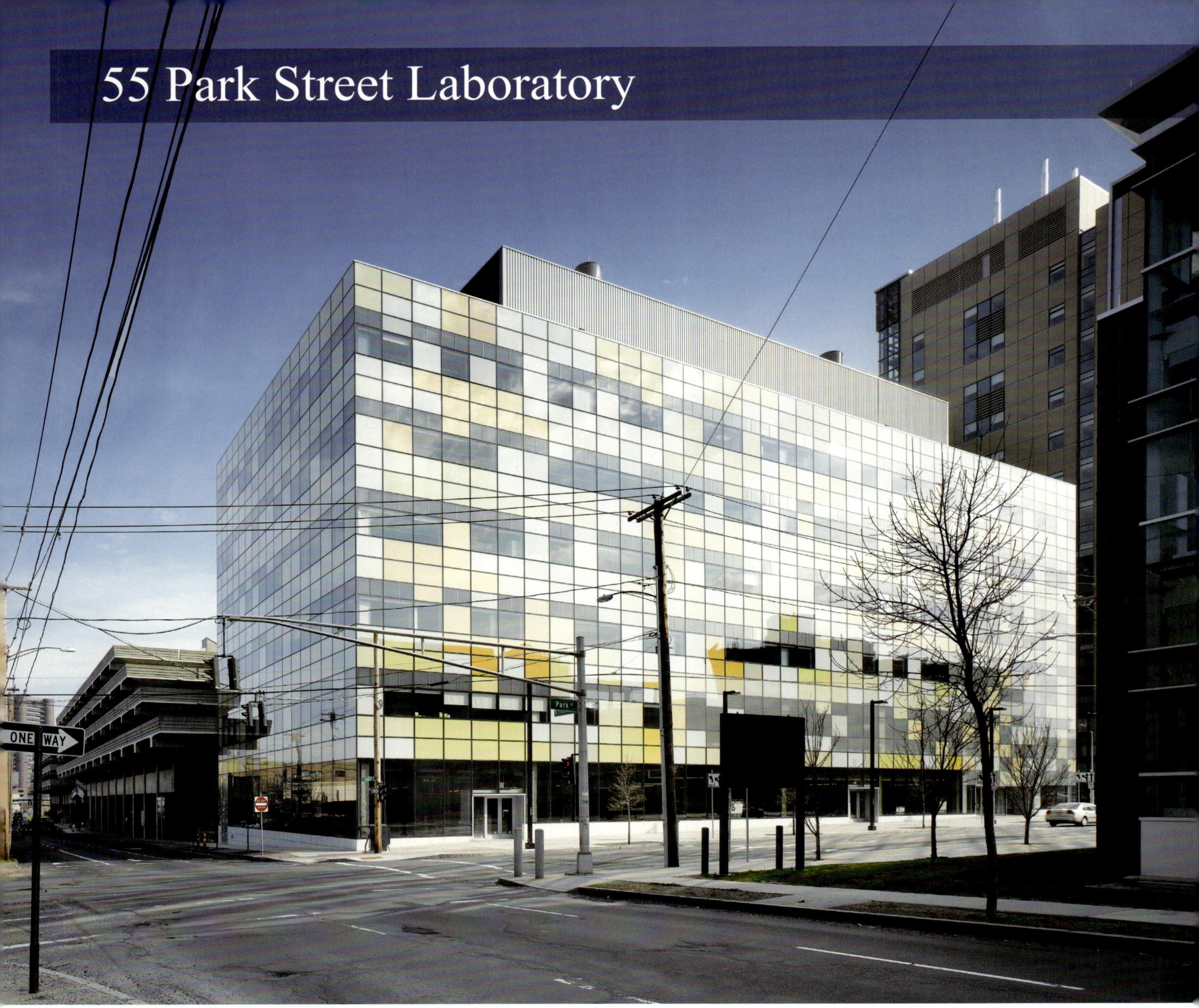

The 55 Park Street Building is a major clinical laboratory extension to the Yale-New Haven Hospital in New Haven, Connecticut; nationally recognized for its commitment to research and teaching. The site is located in the centre of the city, at the threshold between the University and the Hospital / School of Medicine. This part of the city is characterized by the divide created by the Route 34 Corridor, an incomplete urban highway project, and the distinctive, two-storey Air Rights Parking Garage. To date, several master plan studies have attempted to resolve the obvious problems associated with this divide. The 55 Park Street Laboratory is the first building which serves to link the two halves of the city and take full advantage of the amenity provided by the Air Rights Parking Garage.

The original project scope foresaw a stand-alone, multi-storey, clinical laboratory providing necessary support to the neighboring North Pavilion Cancer Center. However, it quickly became evident that the project had the potential to serve as "gateway" to the new cancer center. Program uses of the 55 Park Street Laboratory building include four floors of clinical laboratory space, blood bank, pharmacy, primary loading bay, administrative offices, ground floor retail, and hospital auditorium space. Within the confines of the rectilinear laboratory building the architects have created a large five-storey atrium which serves as a public winter garden; affording researchers, medical staff, patients and visitors an inviting place for all to enjoy in a collective and friendly atmosphere. The winter garden also provides a clear point of orientation on a busy city corner.

The steel-framed building is clad in a high-performance curtain wall system combining highly insulated panels with either clear, colored silk-screened or opaque glazing. The mosaic-type façade pattern on the edifice, with its opaque and transparent panels, filters the natural light coming in on three sides of the building's exterior envelope. The light, together with the imbedded colors and the interior gardens, quietly animates the atrium space, thereby establishing a deliberate contrast to the busy street life outside.

DESIGN ARCHITECT
Behnisch Architekten

EXECUTIVE ARCHITECT
Svigals + Partner

AREA
13970 m²

PHOTOGRAPHER
Roland Halbe

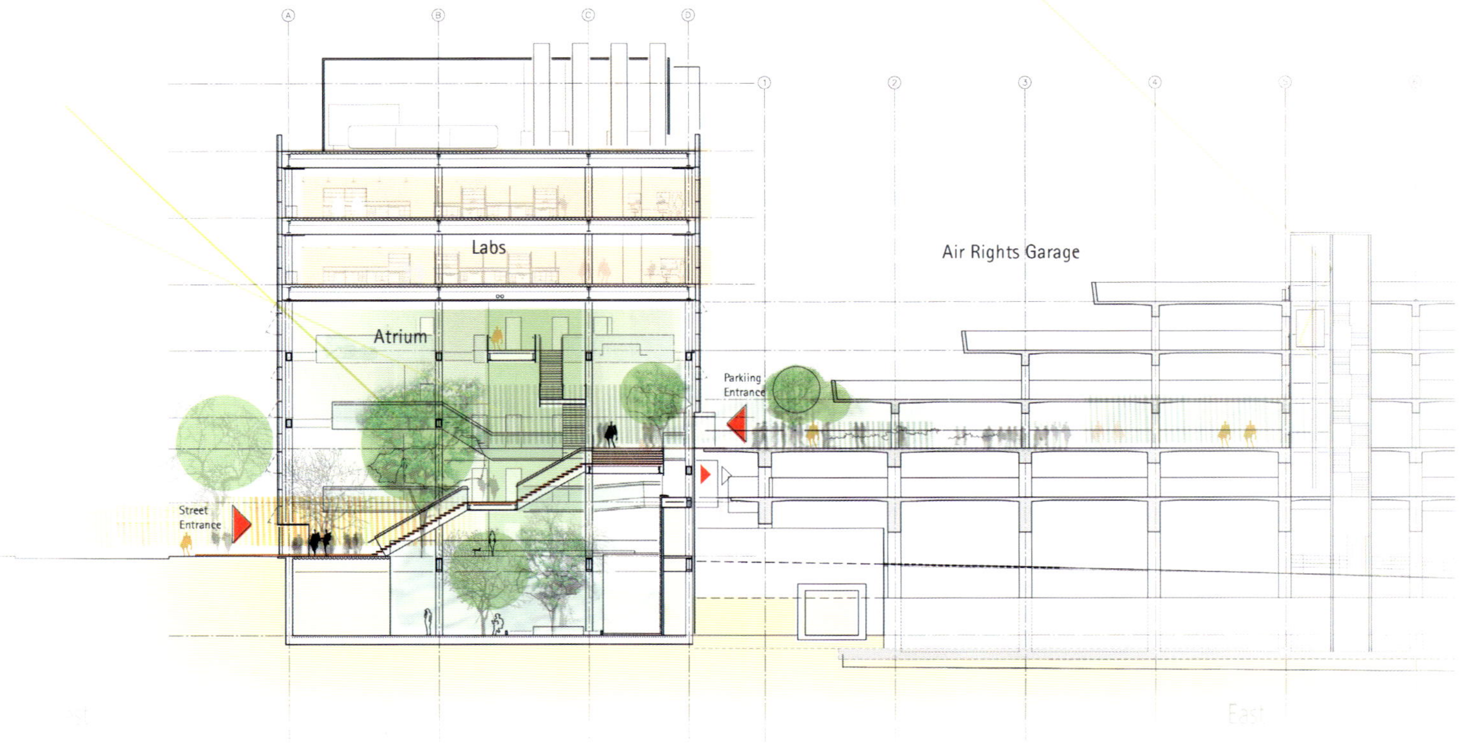
Labs
Atrium
Air Rights Garage
Parking Entrance
Street Entrance

Yale School of Medicine
Yale-New Haven Hospital
Emergency

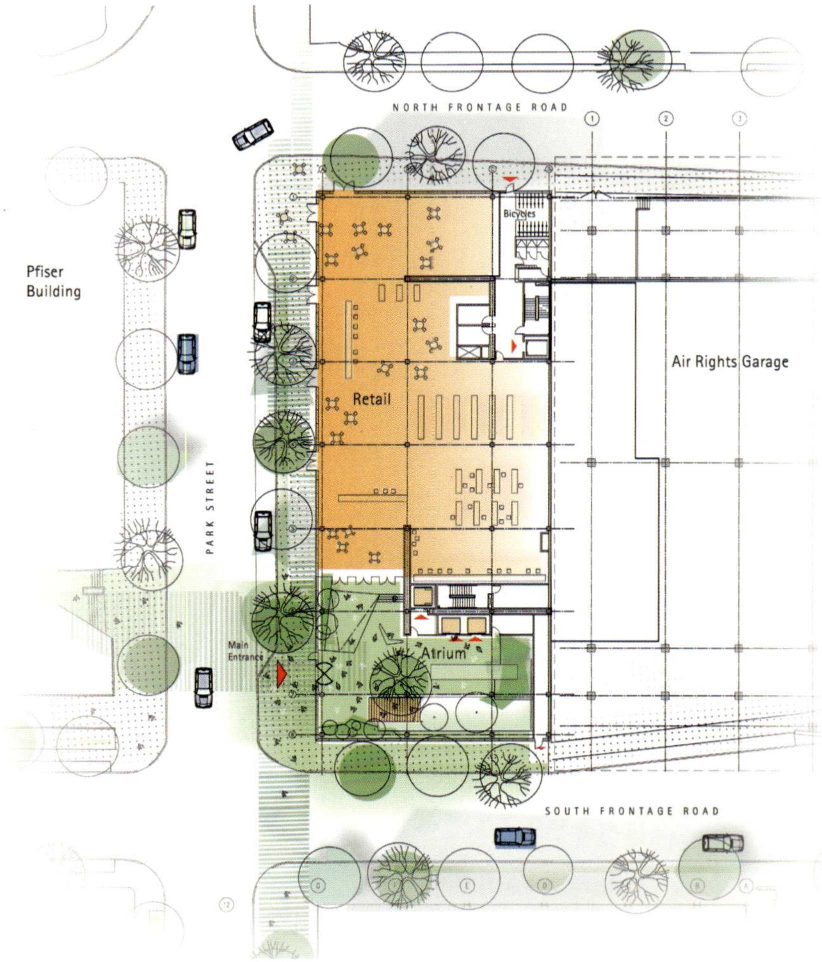
NORTH FRONTAGE ROAD
Pfiser
Building
Bicycles
Air Rights Garage
Retail
PARK STREET
Main
Entrance
Atrium
SOUTH FRONTAGE ROAD

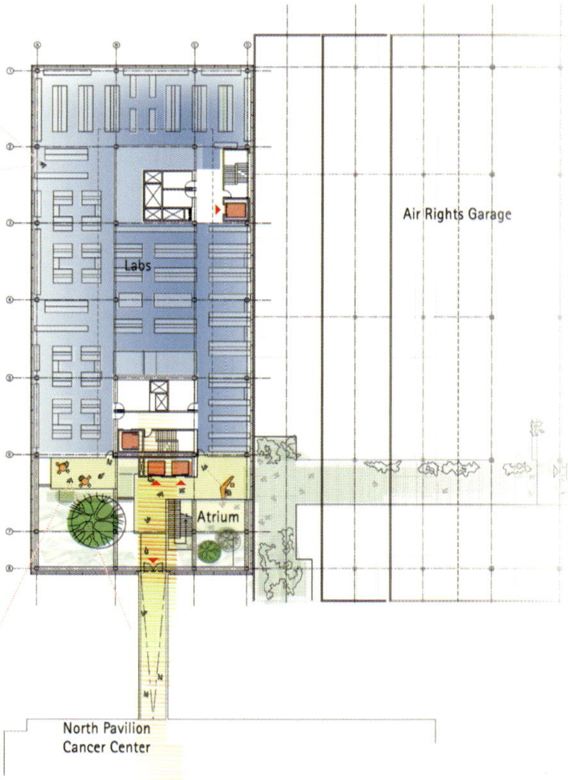
Air Rights Garage
Labs
Atrium
North Pavilion
Cancer Center

TO
91
95

EXIT

公园街 55 号大楼是位于康涅狄格州纽黑文市的耶鲁－纽黑文医院扩建的临床检验科大楼，因致力于教学和科研而闻名全国。该大楼位于市中心，在耶鲁大学与医院 / 医学院交界的入口处。该大楼所在的城区是一个分界线，这个分界线是由没有完工的市政公路工程——34 号公路与独特的双层空中停车场形成的，该分界线将整座城市一分为二。时至今日，仍有城市总体规划试图解决分界线问题以使整座城市融为一体。位于分界线上的公园街 55 号大楼率先将城市分离的两个部分连接起来并充分利用空中停车场提供的种种便利条件。

最初该大楼只计划设计成一幢独立的、多层的、为临近的北楼肿瘤中心提供种种必要配套服务的临床检验科大楼。然而建成之后，该项目很快就显示出作为新的肿瘤中心的"纽带"的潜质。新的肿瘤中心规划使用该大楼的四个楼层作为临床检验科空间，此外还在该大楼建立血库、药房、主装货间、行政办公室、底层零售店和医院的大礼堂。设计师在这座直线形的大楼里设计了一个巨型的五层楼高的天井作为公共的冬景花园，给研究人员、医护人员、病人和来访者营造了一个温馨的场所，让他们在这幢大楼里感受到一种大家互相关联的友爱气氛；同时冬景花园也能让人清晰地找到各部门的位置。

该大楼为钢架结构，外墙为高性能的幕墙系统，将高度绝缘的面板集成在清晰的彩色透明玻璃丝网或者不透明玻璃里。不透明的和透明的面板交替出现构成了大楼马赛克式的墙面，这种墙面过滤了从三个方向透过大楼外立面的自然光。被过滤过的柔和光线、墙面内嵌的各种色彩和楼内花园一起悄悄地让天井空间有了勃勃生机，与楼外喧嚣的街景形成鲜明的对照。

Hungarian Autoklub Headquarters

Vikár & Lukács Architects Studio won the commission for this project on a competition for a new iconic headquarter of the Hungarian Autoklub. The main gesture of the building is a ribbon that wraps around the office spaces on seven floors, while articulating a letter "a". As the site is located near the Danube bridge, local politicians appreciated to have a building that works as a point of orientation for drivers that arrive at district Ujpest.

The main hall and service functions are found on the ground floor, while on the very top, under the arc is a rooftop terrace for representative meetings. The looping strip is about one meter thick, it has a changing width and deforms as it articulates the loop. The two components of the building come back also in the material usage, the strip without openings with metal cladding, and the glass curtain wall of the wrapped building volume. Using geothermal energy system in the building helped the architects to free the rooftop from technical devices.

ARCHITECT
András Vikár, István Lukács, Balázs Domokos, András Klopp

FIRM
Vikar&Lukacs Architects Studio

AREA
3430 m^2

PHOTOGRAPHER
Tamás Bujnovszky, Civertan, Szandra Váraljai, Gábor F. Tóth

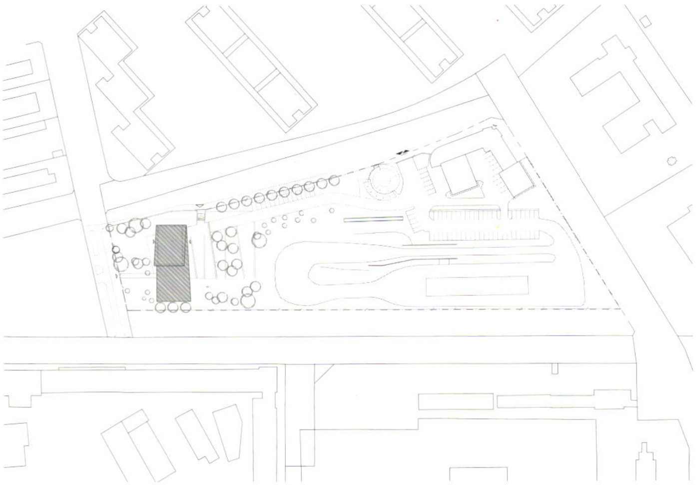

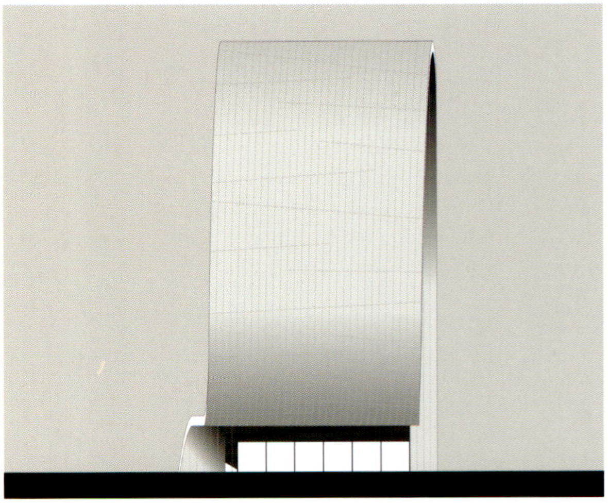

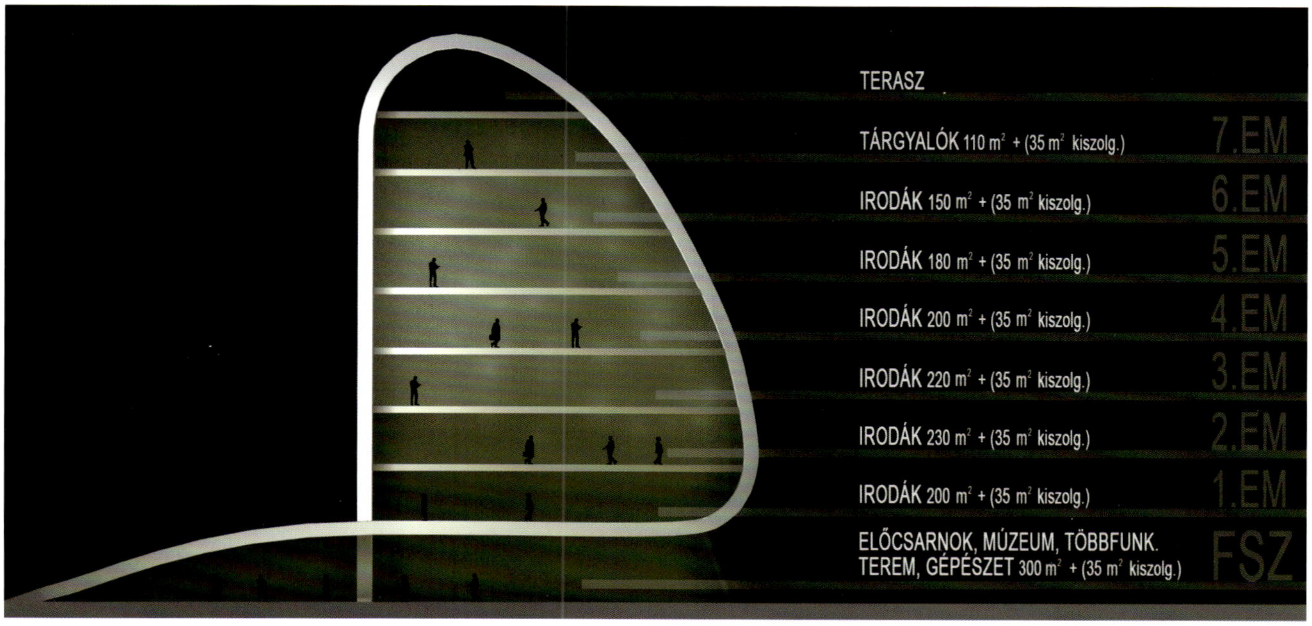
TERASZ
TÁRGYALÓK 110 m² + (35 m² kiszolg.)
7.EM
IRODÁK 150 m² + (35 m² kiszolg.)
6.EM
IRODÁK 180 m² + (35 m² kiszolg.)
5.EM
IRODÁK 200 m² + (35 m² kiszolg.)
4.EM
IRODÁK 220 m² + (35 m² kiszolg.)
3.EM
IRODÁK 230 m² + (35 m² kiszolg.)
2.EM
IRODÁK 200 m² + (35 m² kiszolg.)
1.EM
ELŐCSARNOK, MÚZEUM, TÖBBFUNK.
TEREM, GÉPÉSZET 300 m² + (35 m² kiszolg.)
FSZ

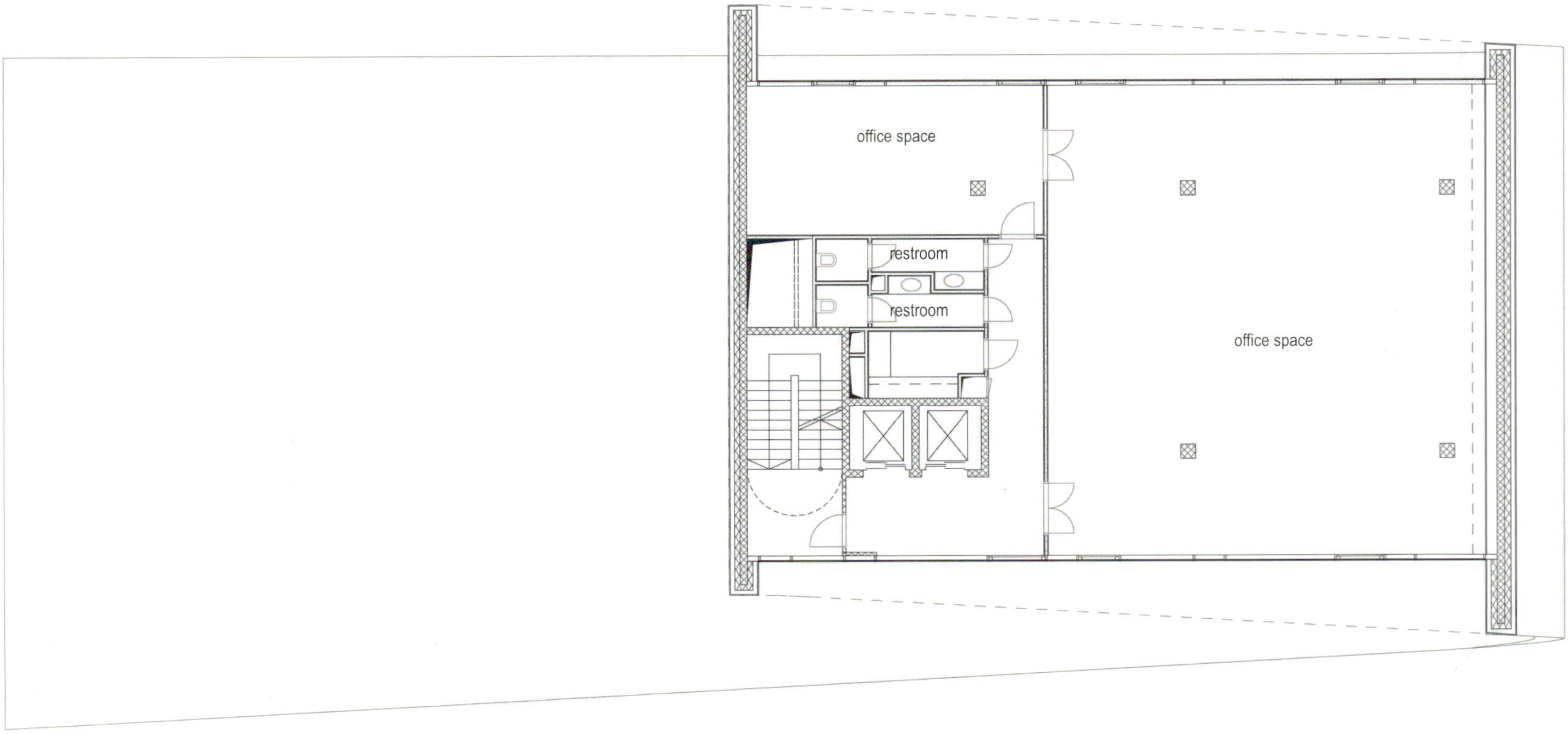
office space
restroom
restroom
office space

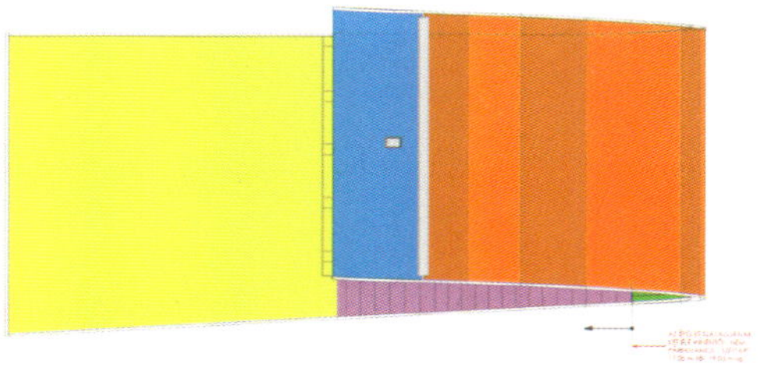

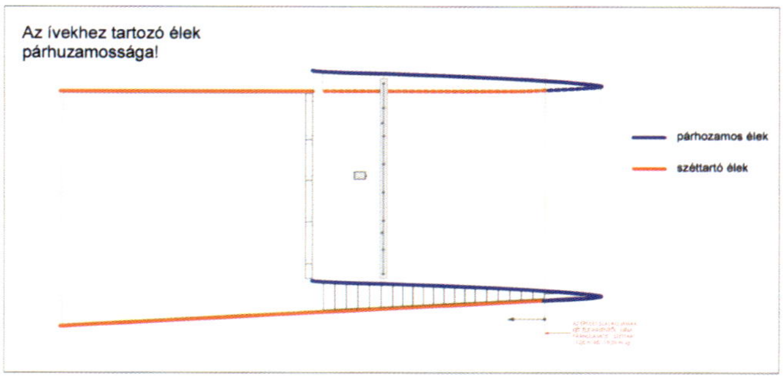
Az ívekhez tartozó élek
párhuzamossága!
párhozamos élek
széttartó élek

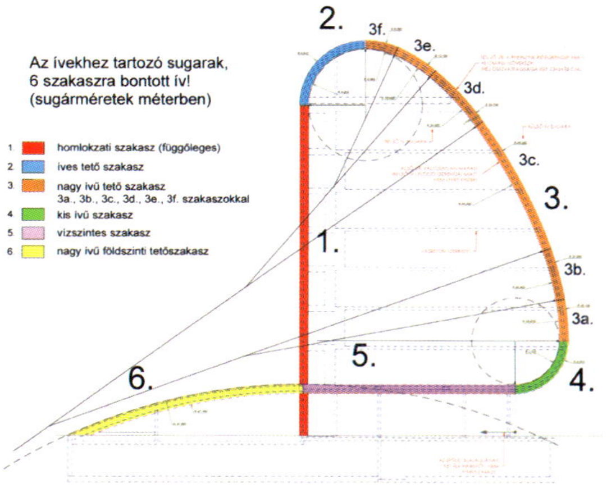
Az ívekhez tartozó sugarak,
6 szakaszra bontott ív!
(sugárméretek méterben)
1. homlokzati szakasz (függőleges)
2. íves tető szakasz
3. nagy ívű tető szakasz
3a., 3b., 3c., 3d., 3e., 3f. szakaszokkal
4. kis ívű szakasz
5. vízszintes szakasz
6. nagy ívű földszinti tetőszakasz
1.
2.
3.
3a.
3b.
3c.
3d.
3e.
3f.
4.
5.
6.

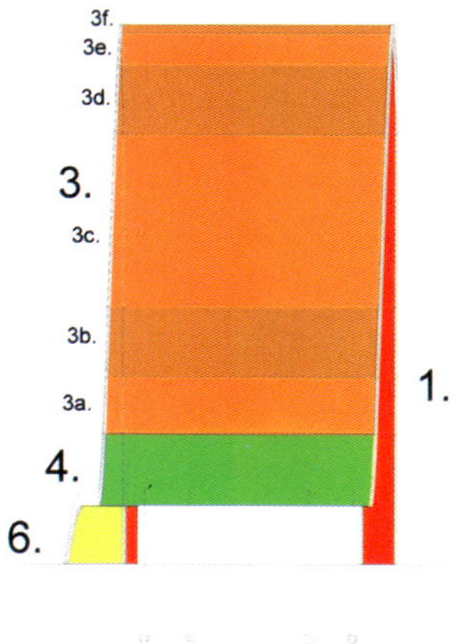
3f.
3e.
3d.
3.
3c.
3b.
3a.
4.
6.
1.

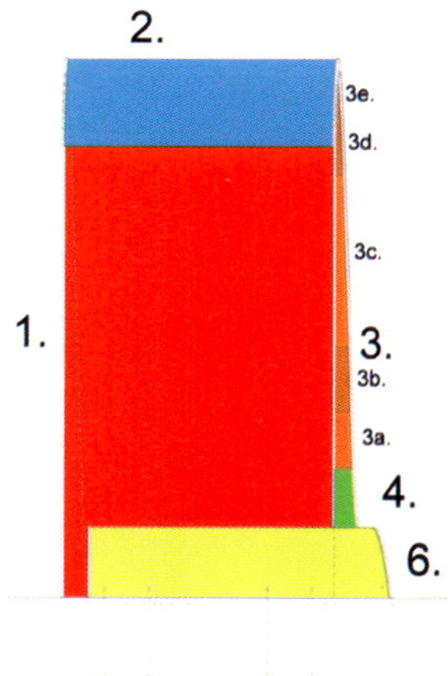
2.
3e.
3d.
3c.
1.
3.
3b.
3a.
4.
6.

Vikar & Lukacs 建筑设计事务所在匈牙利 Autoklub 总部设计竞赛中赢得并完成了 Autoklub 总部的设计工程。该建筑最主要的亮点是如同缎带一样的外观，包裹住七层的办公建筑空间，醒目地写下一个巨大的字母“a”。建筑紧邻多瑙河桥边，当地政客希望将该大厦建成重要的地标性建筑，帮助进入 Ujpest 区的司机定向。

大楼一层设有主厅和服务功能区。大楼顶层弧形楼顶之下是屋顶露台，设为会议室。包裹整个建筑结构的环形缎带厚约一米，其宽度和形状随着环绕的需要不断变化。大厦的另两个设计元素体现在材料使用方面：大厦的带状体为金属覆面，整个带状体没有任何开口；被包裹的整个办公空间为玻璃覆面。整个建筑利用地热系统供暖与制冷，避免了在大厦顶端装载任何技术设备。

Elbek & Vejrup

ARCHITECT
CEBRA
AREA
2800 m^2

The Elbek & Vejrup company headquarters is laid out as a traditional square atrium building. It is the first of the two office buildings that will eventually share this plot on the outskirts of Aarhus. The plan is organized in a diagonal system resulting in rotational symmetry with stairs, cores and entrances placed in opposite corners.

As Elbek & Vejrup develops software systems for a wide range of clients, the architects were asked to reflect reliability and digital high-tech in the building. The architectural idea is based on the obvious link between brick and pixels. Just as a picture or video is made up of thousands of small squares of color, a house is built of bricks or building blocks. The architects chose to work with bricks in a straight grid bond to give an old tried and trusted building technique a digital twist. The idea of pixels runs through the entire project like an obsessive compulsive disorder. The tiles in the restrooms are decorated with 8-bit space invaders and the custom-made carpet tiles spell the company name in binary numbers built from pixels. Even the green plant walls are divided so the different species form giant pixels.

The choice of brick has a symbolic cause, but it also supports the company as a reliable business partner. Bricks are sturdy and they express the quality and timelessness, but when, as here used imaginatively, you get just as much to think about innovation. In short, the architects seek through the architecture to portray a balance between reliability and digital creativity – between brick and pixels.

LOCATION\Aarhus, Denmark

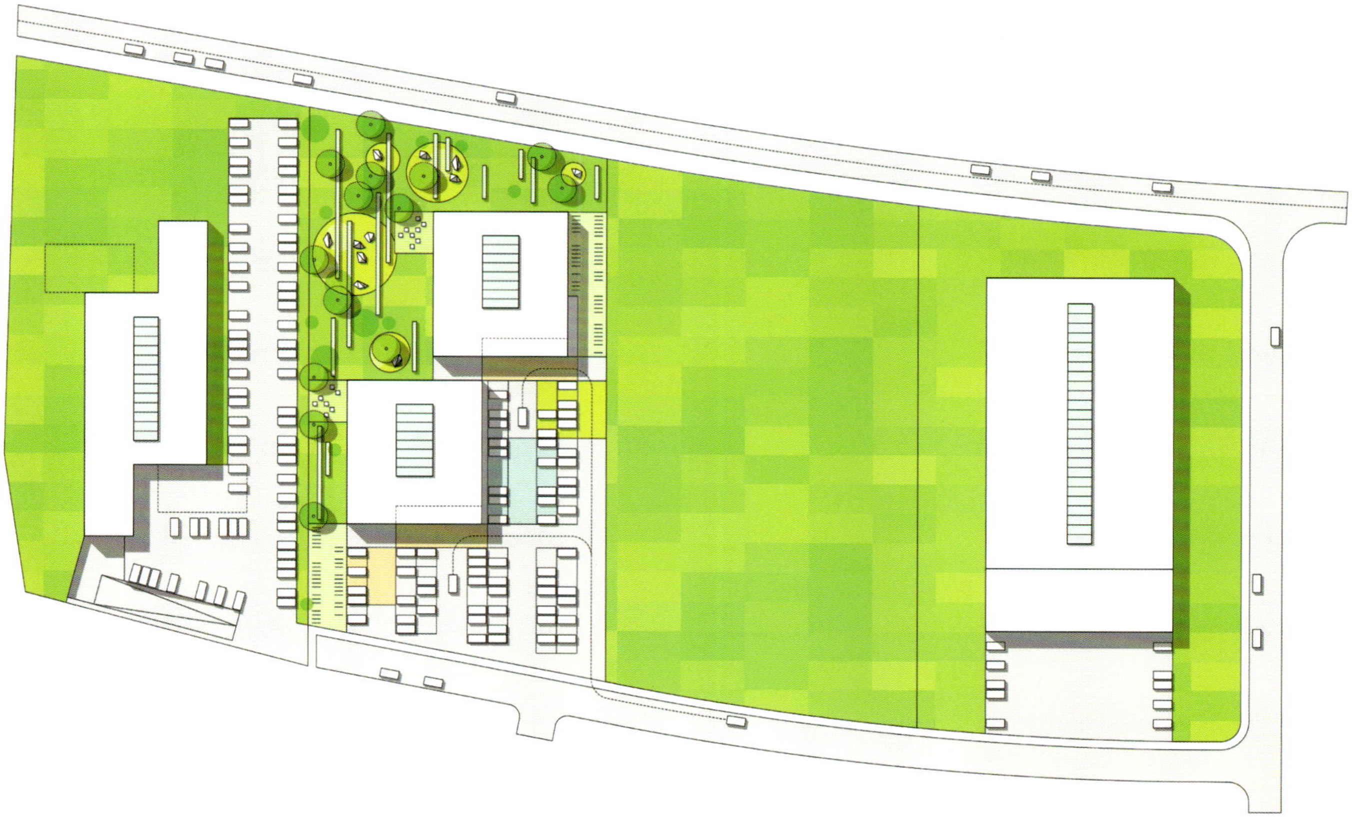

briXels
2009
M.F.

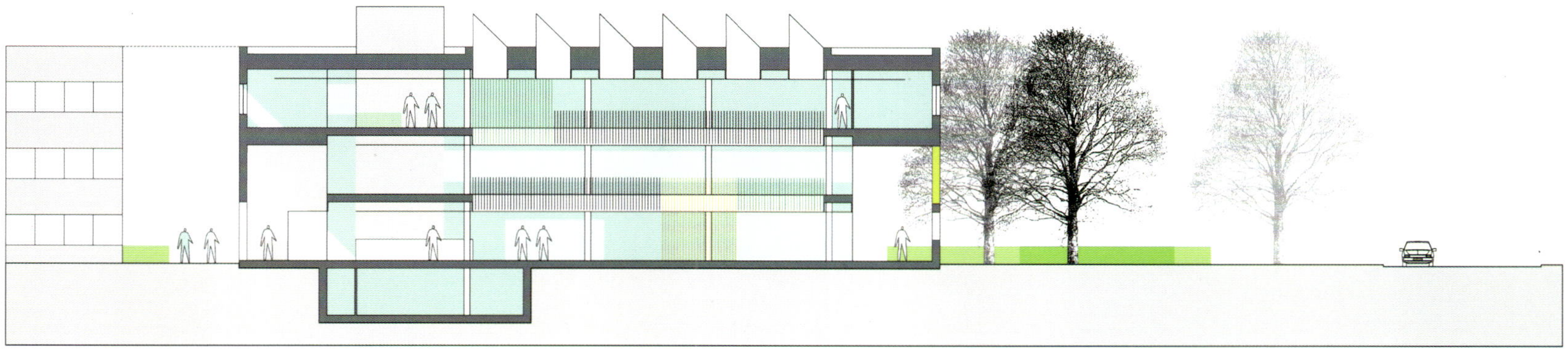

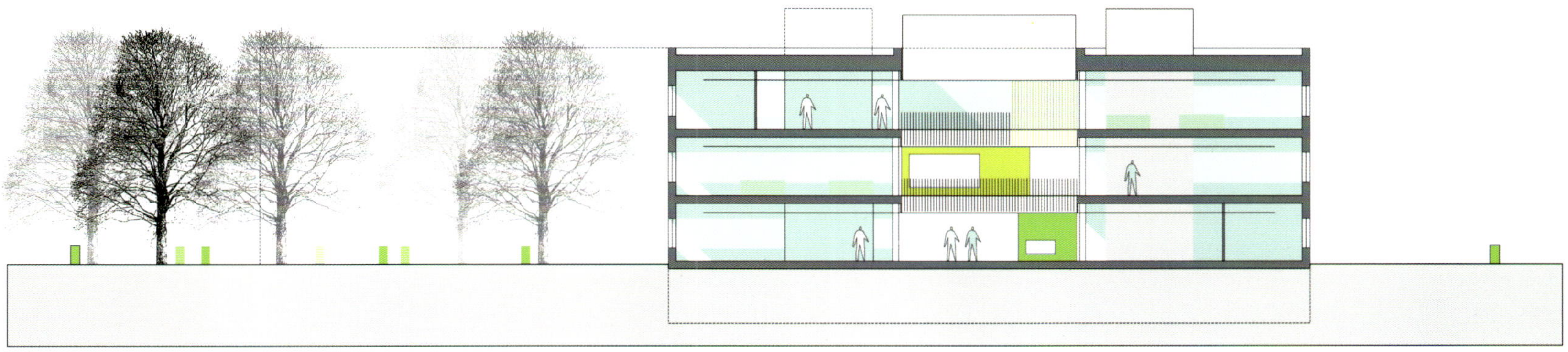

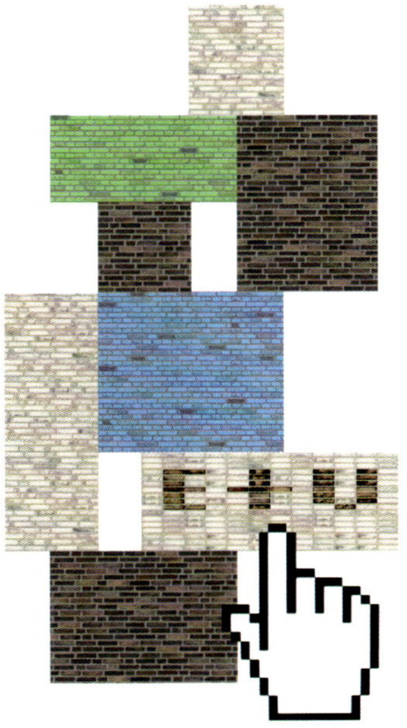

Elbek & Vejrup 公司总部大楼设计成了一座传统的具有正方形天庭的建筑。在 Aarhus 市郊区的一小块地皮上最终将建有两座办公大楼，而这座大楼是最先建起来的。整个工程规划遵从对角线策略，因而相对两座大楼的楼梯、中心区域以及入口对称分布。

由于 Elbek & Vejrup 公司为不同行业的众多客户开发各种软件系统，所以公司要求建筑师们在建造大楼时要通过建筑充分反映出公司的可靠性和数字化高科技理念。Brixels 的建筑理念是基于砖块与像素之间的关联，就如一幅画或者一段视频是由成千上万块颜色块所组成的一样，一座房子也是由成千上万块砖块或者建筑构件建成。建筑师们选择的方案改变了古老的一成不变的建筑技巧，而将砖块砌成连接的网格状，使整个外墙呈现出数字化的形象。这种像素理念贯穿于整个工程的设计中：卫生间里的瓷砖装成八位字节空间的形状；专门定制的塑钢方块毯设计成像素状二进制数字，在二进制数字中拼出了公司的名字；甚至绿色植物装饰的墙面也由不同的植物品种分隔开来，并形成巨型像素状。

砖块的使用方案除了具有象征性的目的外，同时也树立了公司作为可靠的商业伙伴的形象。砖块质地坚固，象征着品质高，使用时间长，充分利用想象，甚至可以由此联想到创新。简言之，建筑师们希望通过建筑本身，通过像素和砖块之间的平衡来传递一种可靠性和数字型创新性之间的平衡。

Dogan Media Center Ankara

The building is strongly perceived from afar and in diverse prospects at various angles due to perforated shield resembling Braille alphabet at range of sizes, symbolizing "communications for all". The metal panels filter the sunlight entering the building into shifting patterns of dappled shade. The panels' perforations are echoed in the suspended ceilings, where circular cut outs house connections for camera equipment and lighting.

The projecting boxes are each dedicated to a specific TV channel or newspaper so that the various units within the Doğan conglomerate can be readily identified from afar. Intermediate floors are supported on secondary steel columns and beams. There is a layering of views through the building, both horizontally, across the open-plan offices and out into the city, and vertically, in the way the different levels open onto the internal atrium. The floors are framed by parapet walls topped with glazing. Here, the transparency is complete. The parapets consist of frameless glass supported by a stainless steel handrail; above the rail, there is only air.

Separate sections like studios and offices of press people preserve their exclusivity and each segment is accentuated in the integral form of the re-assembled blocks of cube. The assemblage of smaller cubes within the structure of the large box can be read as a metaphor for the diversity of the company's operations and a high degree of transparency serve to reinforce the separate brand images.

The top of the building contains a VIP lounge and terrace with teak decking. The floors below are segregated according to their specialty – newspaper or TV – but everyone comes together in the canteen in the ground-floor atrium. There is an open car park in an L-form parallel to the building besides two underground levels reserved for parking. The first basement is saved for technical facilities and storage requirements are solved in basement floors. The first basement level contains large studios as well as support facilities such as a hairdresser and make-up rooms. The two floors below contain M & E equipment and parking, along with cisterns that allow the landscaped gardens to be irrigated with grey water. Technological facilities and healthy infrastructure makes the building user-friendly and provides easy maintenance.

ARCHITECT
Tabanlıoğlu Architects, Melkan Gürsel – Murat Tabanlıoğlu
ARCHITECTURAL PROJECT TEAM
Murat Cengiz, Çağrı Akay, Ozan Öztepe, Ali Eray
AREA
11475 m^2
PHOTOGRAPHER
Cemal Emden

LOCATION\Ankara, Turkey

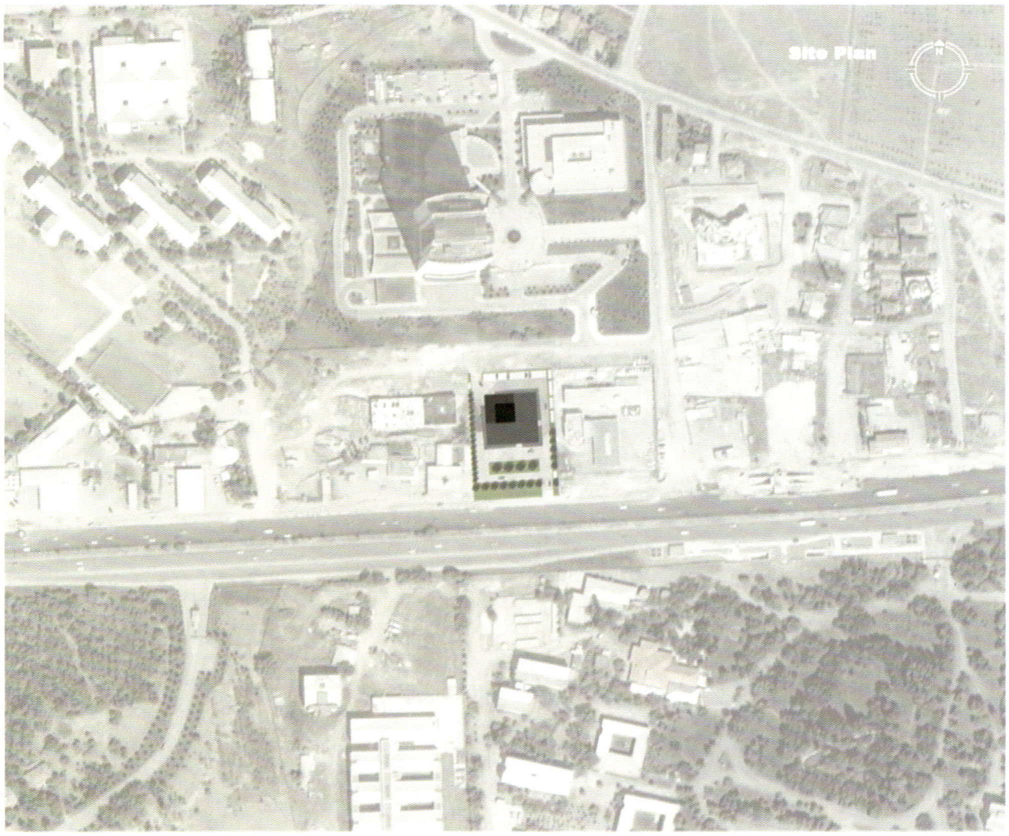

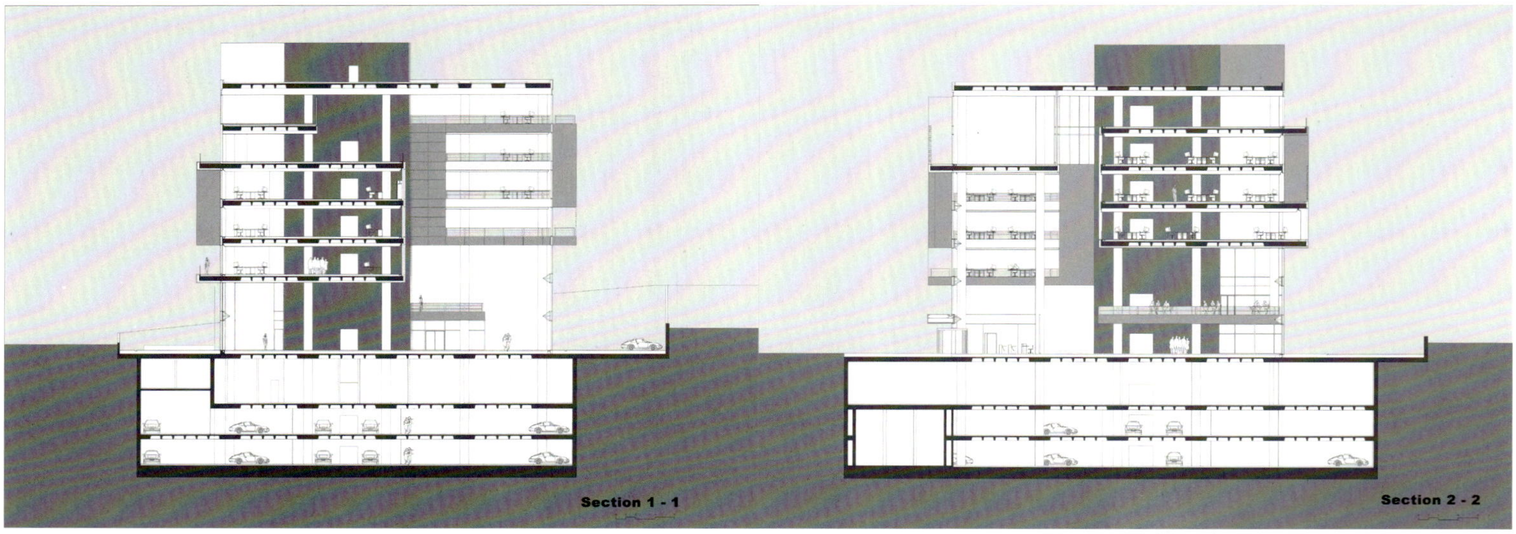

位于安哥拉的 Dogan 媒体中心大楼从远处不同的角度看都会感受到视觉的冲击，因为大楼立面上的穿孔防护板看起来就如大小不一的盲文字母，象征着“与所有人沟通”的意愿。阳光照射到带孔的遮阳板上，经过遮阳板的过滤，斑驳的阳光才照进大楼。室内垂吊式天花板上的孔眼与冲孔立面遥相呼应，孔眼处的环形凹槽设置有连接线，以备连接摄像和照明设备。

从表面看，一个个突出的方形体，每一个都容纳了不同的电视频道或报纸，这种布局使人们从远处就能看出 Dogan 集团内部的各个单位所在的楼体位置。中级钢柱和钢梁支撑着中间楼层。站在大楼不同楼层所看到的景致都是不同的。在水平方向，透过开放式的办公室可以看到远处城市的风景；在竖直方向，可以欣赏到内庭的美景。楼层的四周安装着护栏，护栏的最上端是玻璃的，形成完全透明的效果。护栏本身由不锈钢扶手与无框玻璃组成。

每一个独立的单位，例如，影音工作室和新闻工作者办公室的装置保持了自己的独立性和独特性，每一个独立单元都强调了楼体内部空间的独特装配。在整个大楼结构里，规划设计有一个个独立的小区间象征着公司多样化的运作理念和手段，楼体的高透明度也是为了增强内部各个单位的品牌形象。

大楼顶层是装有柚木露台的贵宾厅。较低各楼层根据其所容单位专业——报纸或者电视——的不同而分隔在不同的空间。一楼中庭里设有餐饮部，也是大楼内的员工就餐和聚集的场所。除了设置于地下的两个停车场之外，还设置有一个平行于建筑的 L 形露天停车场。技术设备和其他材料存储在地下一层，这里还设有大型的工作室和一些配套辅助设施，如美发室和装配室。地下二层装有机电设备和停车场，还有用于储备盥洗污水以灌溉园林的水箱。大楼内的技术设施和保健设施极大地方便了用户，也易于保养。

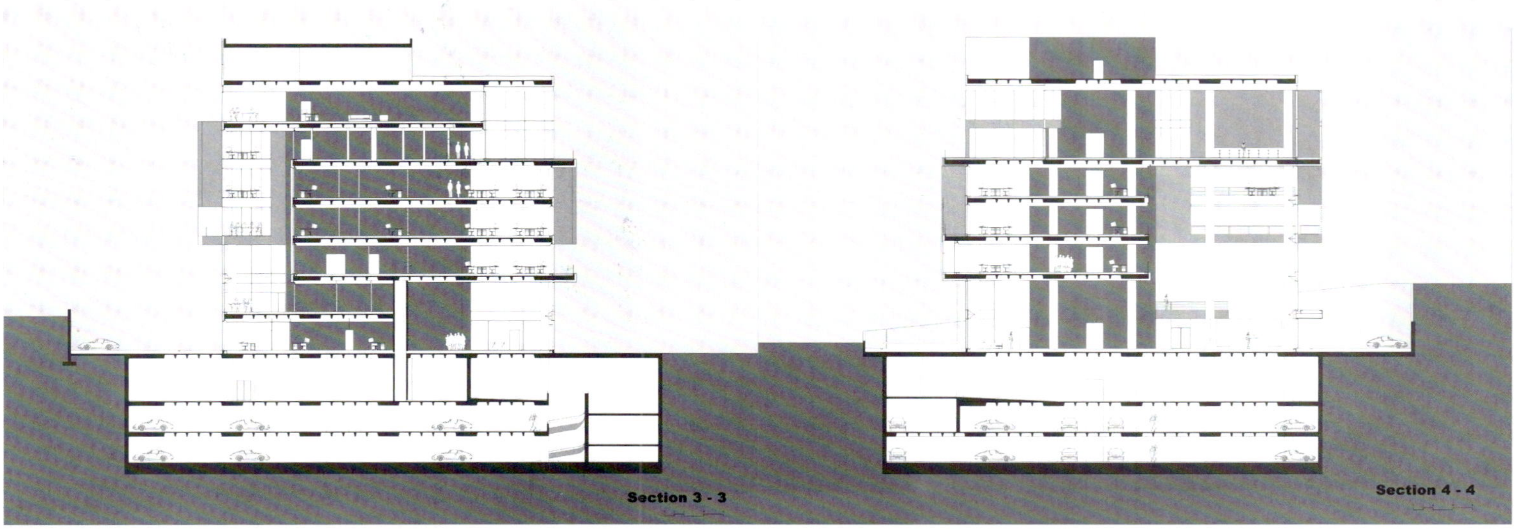
Section 3 - 3
Section 4 - 4

VIP Plan

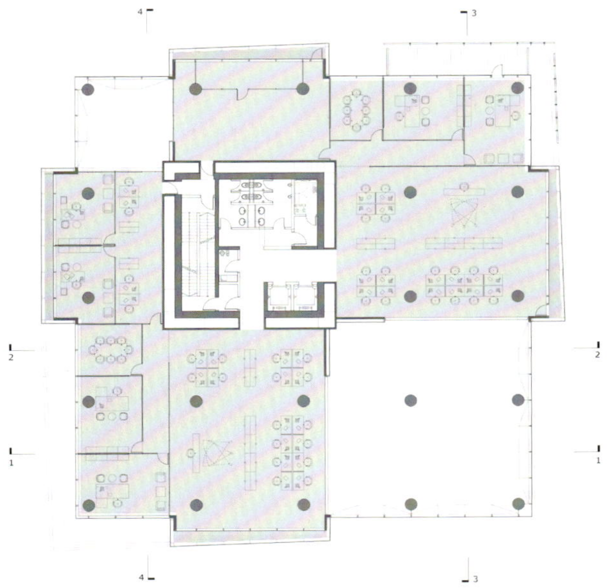

4th Floor Plan

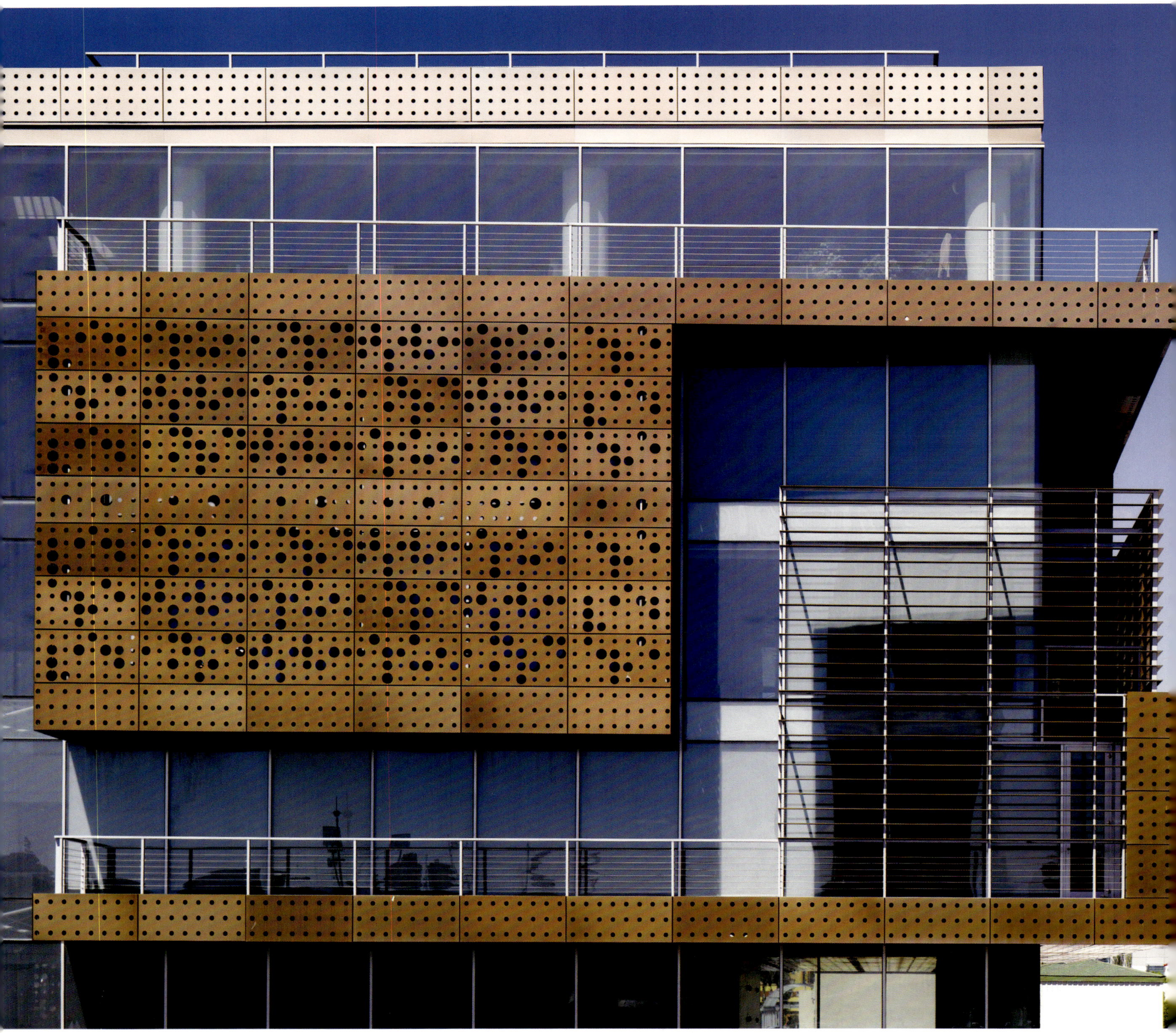

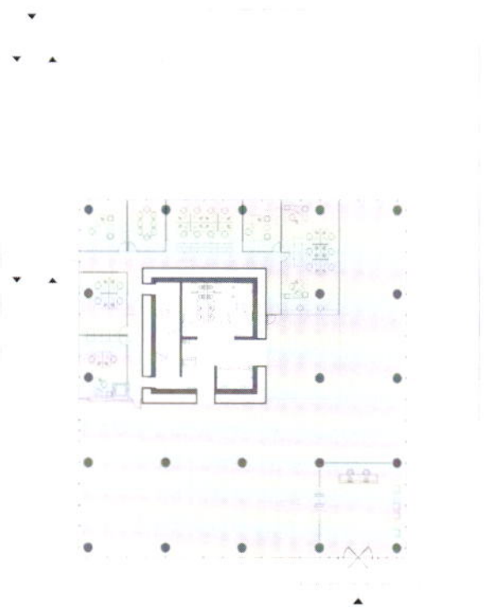

Ground Floor Plan

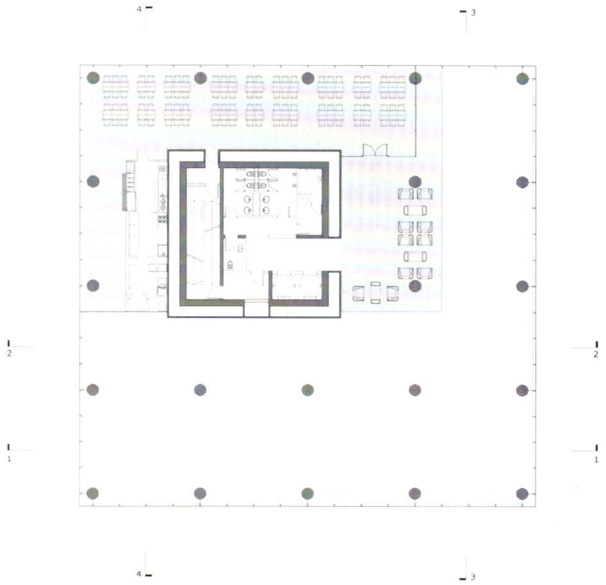

First Floor Plan

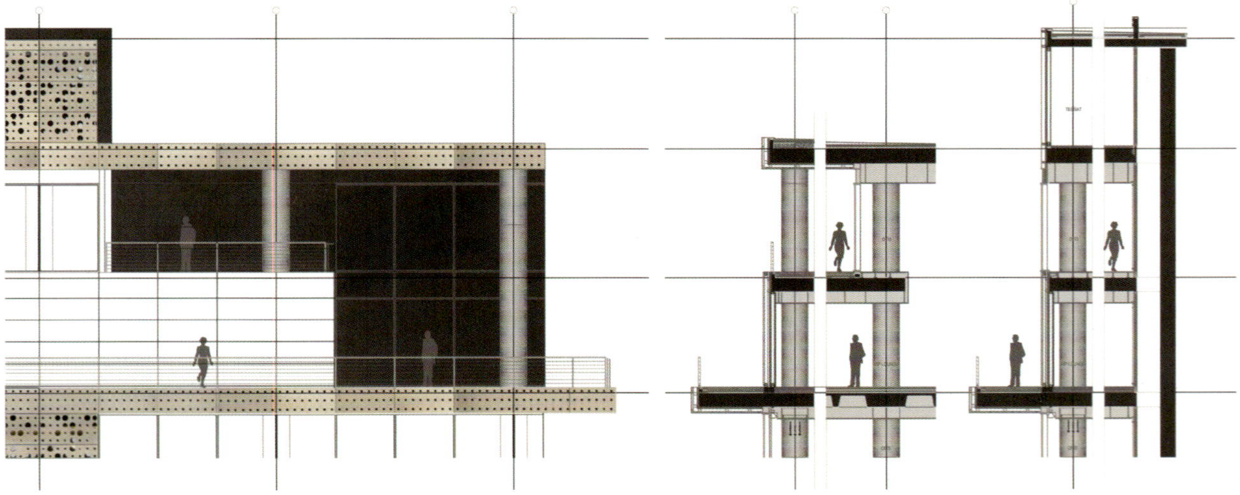

System Detail

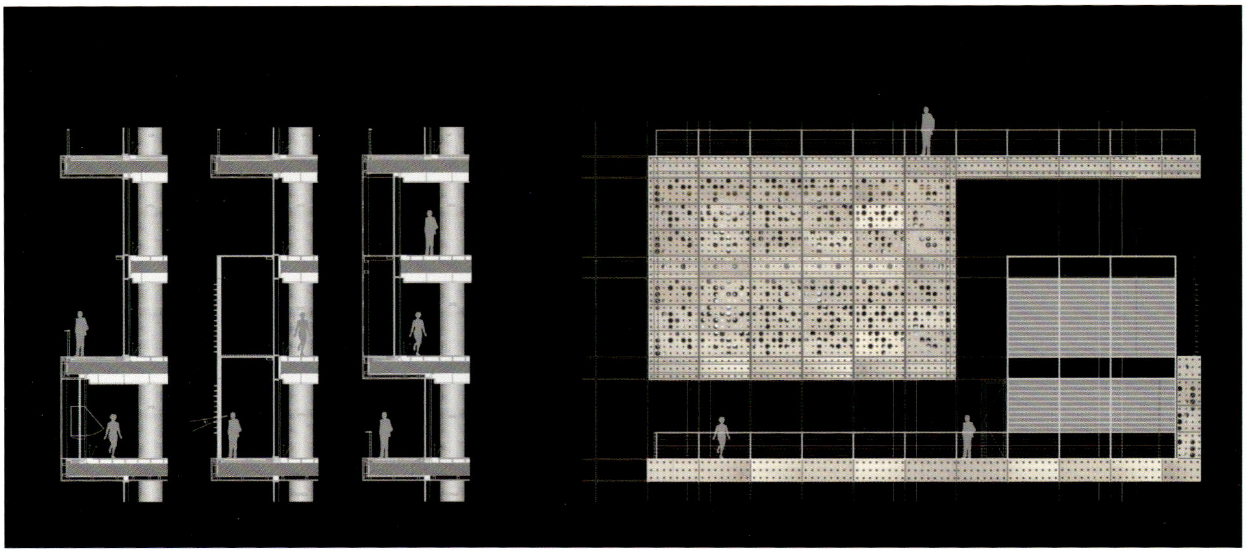

Wilkie D. Ferguson Jr. United States Courthouse

The building stands at the end of the axis created by the 4th Street Promenade. However, because the center atrium space is approximately 20 meters above the plaza, both the physical and visual axes are maintained through the building. The design completes the judicial campus created by the old courthouse, the courthouse annex tower, the Lawrence King Building and the federal prison.

The planimetric design is based on a four-courtroom per floor layout. The two court towers are connected by a generous circulation lobby, which occurs on every floor, and rises dramatically throughout the entire height of the building. A dramatic conical glass atrium connects the floors by piercing through the lobbies and decreasing in size as it rises through the building terminating in a skylight.

The site plan is divided into two distinct zones. The first, at street level is comprised of densely planted native vegetation punctuated with seating and meandering walks which can be used for local art fairs and markets. The second zone is the lawn. It is raised approximately 1 meter above the sidewalk forming a natural and secure perimeter around the building. The two east quadrants of lawn areas were designated as the sites in which the artist Maya Lin developed her earthen installation. The hard-scape is primarily comprised of three-colored pre-cast pavers, who coordinate, with the pre-cast panels of the building.

The building is composed of three individual elements: the opposing two towers and the glass crystal that spans between them. The east and west curtain-wall elevations of the two towers are a blueprint for the interior functions within the building. The alternating rhythms, depths and colors of the horizontal and vertical sunshades delineate the primary office, circulation, chamber and courtroom functions located in the tower. The north and the south facades of the tower are solid pre-cast with punched openings reinforcing the solidity of the envelope which then create the asymmetrical frame on the east and west elevations.

The signage is located on all four elevations of the tower. At the base, it welcomes pedestrian, at the top it acts as a banner for the vehicular traveler. The exterior expression is primarily transparent and open reflecting the integrity of the court while emphasizing equality before the law. A monumental three color stone colonnade recalling the richness of traditional courthouses supports the south tower. The north tower sits upon the building garage clad in stone columns placed at irregular intervals and a metal grille. The crystal curtain-wall spanning the two towers contains an atrium that pierces through each floor.

ARCHITECT
Bernardo Fort Brescia, Laurinda Spear

FIRM
Arquitectonica

ASSOCIATE ARCHITECT
Hellmuth, Obata + Kassabaum

AREA
53700 m^2

PHOTOGRAPHER
Robin Hill, Norman McGrath

SOUTH ELEVATION

EAST ELEVATION

威尔基·D. 弗格森美国迈阿密联邦法院耸立在第四大街步行广场的中轴线末端。其中庭区域比步行广场高出近20米，因而无论是实际上还是视觉上，该大楼都横跨轴线。这个工程完善了由旧法院、法院的配套建筑即劳伦斯·金大楼和联邦监狱所形成的司法园。

该工程的平面设计是基于每个楼层设置四个审判庭的布局进行的。宽敞大气的流通大厅将两座塔楼连接起来，绕着中庭从下至上与塔楼各层相连。圆锥状的玻璃中庭通过大厅将各层连接起来，由下至上逐渐变小，最终在楼顶变成了天窗。

该项目的总设计规划分为两个不同区域。紧邻街道的区域长着浓密的本土植物，设有坐椅和蜿蜒的小道。这些设施使该区域可用来举办当地的艺术博览会和艺术品集市。另一个区域是草坪区。草坪比人行道要高出一米，很自然地环绕在大楼的周围。草坪区的东边开辟出来，艺术家林璎设计的陶制装置艺术立在那里。硬质地面上铺的是三色预制地板砖，其颜色与大楼预制面板的颜色搭配协调，互相辉映。

大楼由三部分构成：两栋塔楼相对而建，中间的玻璃“水晶”状楼体起到了调和的作用。塔楼东西立面的设计构造反映出大楼内部的功能规划：颜色迥异、深浅不同的水平和纵向遮阳系统富于韵律感，同时也反映出塔楼内的主要办公空间、流通区域、议事室和审判室之间的功能差别。塔楼的南北立面应用了预制石材，窗户设计成单独的冲压式开口，立面因此增加了坚固的感觉，同时也使东西立面的框架结构不完全对称一致。

塔楼四个方位的立面上都安装了标识牌：在底层的标识牌表示了对行人的欢迎，而在顶层安装了让驾车人远远便能看见的巨幅标识，使该建筑无论是在白天还是夜晚，都成为当地的标志性建筑。建筑从整体上呈现透明和开放的形象，其设计初衷是用建筑语言反映出法律面前人人平等，表达法庭的正义使命。南楼的底部是宏伟的三色石制柱廊，烘托出法庭丰富而多彩的历史传统。北楼的底部是车库，间或排列着石柱，并安装有金属格栅。起调和作用的“水晶”状楼体部分设置有天井，天井从下到上贯穿着楼体的每一层。

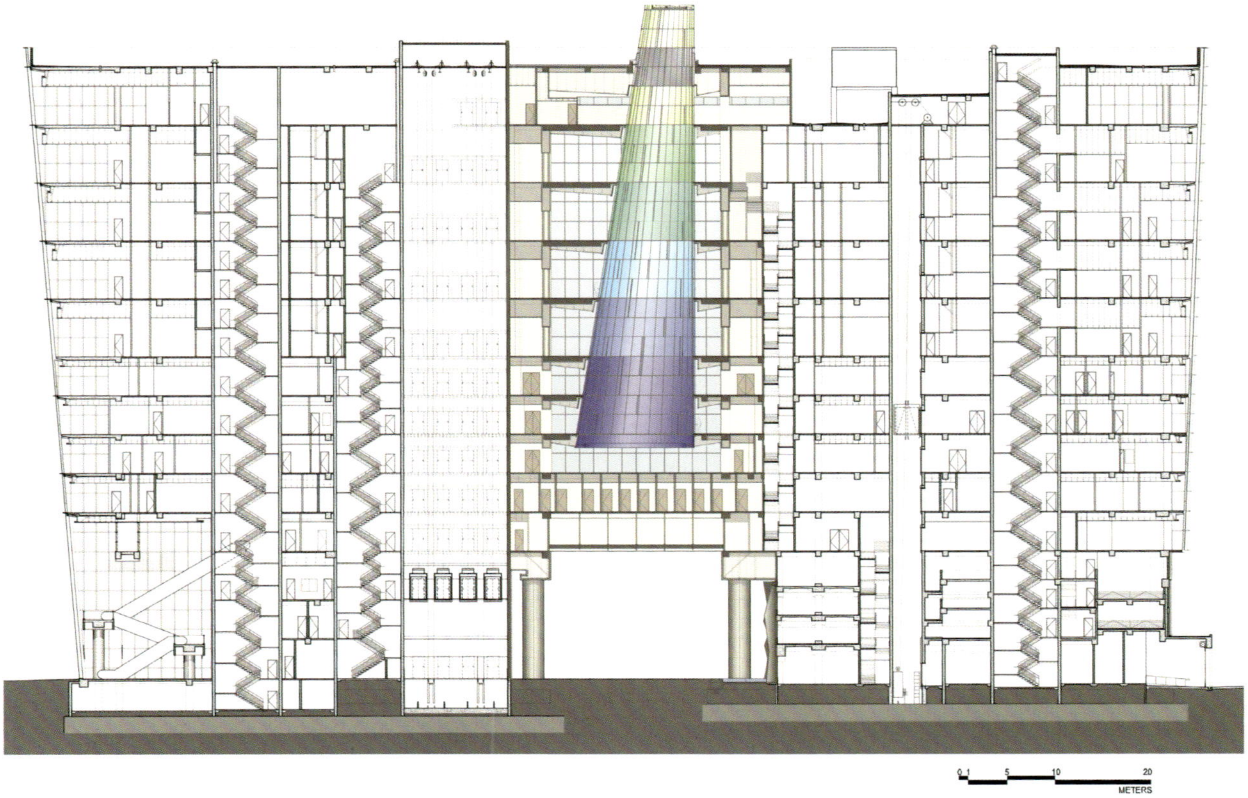
0 1 5 10 20
METERS

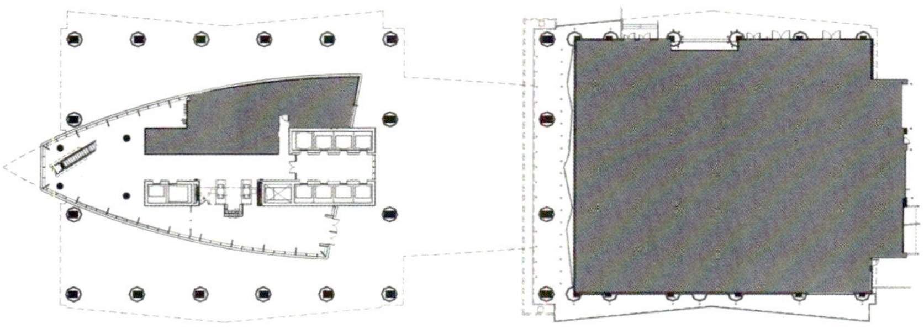

Ground Floor Plan

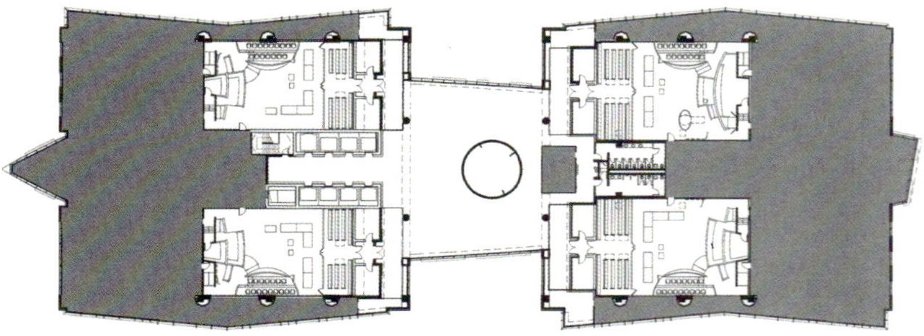

Courts Floor Plan

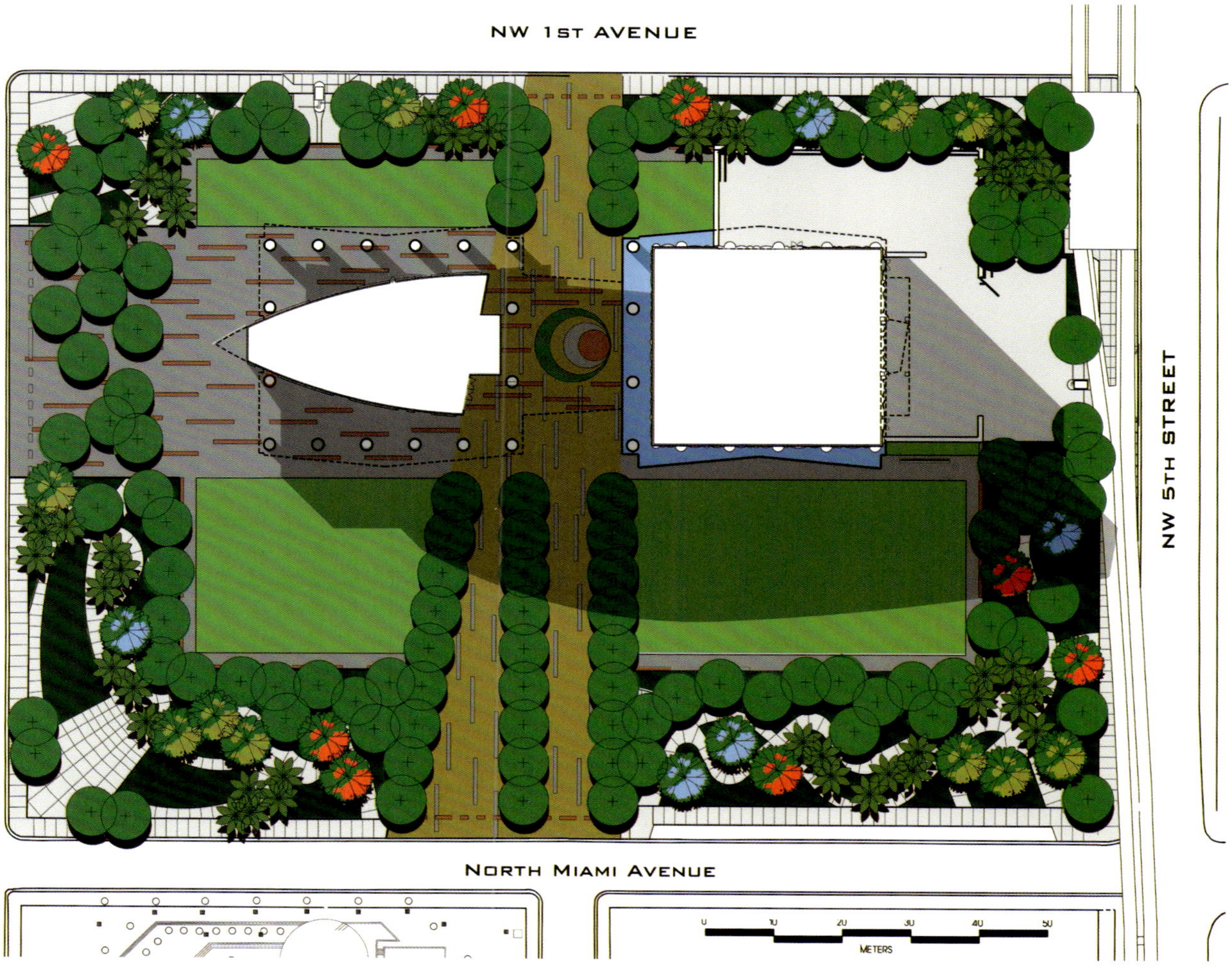
NW 1ST AVENUE
NW 3RD STREET
NW 5TH STREET
NORTH MIAMI AVENUE
0
10
20
30
40
50
METERS

FERGUSON, JR.
UNITED STATES COURTHOUS
E PLURIBUS
UNUM

AIC Automotive Intelligence Center

The complexity of the main building's facilities is unified under a large silver aerodynamically-shaped roof inspired by the bodywork of prototypes and racing cars.

The main building is composed of two fundamental parts: the social area, set around the main lobby; and the area for R&D projects.

In the transverse section of the R&D area, the functional layout is understood in the solution that has been given to the problem of the confidentiality of the activities in the center. Continuing from west to east, we have the garden, a landscaped fold in the terrain generating a patio alongside the façade which we call a "burladero". It is serves as a screen so that small machines can be installed for the laboratories of each company without any visual impact.

Continuing at the ground floor level, an internal access "street" with the laboratories off it, for the companies one side and on the other side those in common. Above it, at the first floor level, a passageway fulfils the same service for people to access the offices of the companies. And finally the area of the shared laboratory, the highest enclosed space in the building. The lower parts of the roof are reserved for future "units" of the shared projects. At the end, the East façade is designed like a fold of the roof down to the ground, negating the presence of the combined cycle power station.

The "street" is covered in order to reinforce the image of unity the centre wishes to convey; not to mention the advantages in standing up to the Vizcaya climate, as well as facilitating the control of some habitually untidy spaces.

The social area is planned at the head of the building. A large lobby facing south forms the end of the internal "street". Some huge glass doors make this connection possible. The training areas are integrated into an aluminium box: an auditorium downstairs, training classrooms upstairs.

The management is located in the most aerodynamic element of the roof: a fold inspired by the air intakes of racing cars. It facilitates a generous access porch and gives the centre's management to have a commanding view of all the facades.

Urban Insertion

The complex is located on an industrial estate away from the urban nucleus. The design strategies have been geared, on the one hand, to making the building visible from the bordering motorway, and on the other hand, to avoiding the constant view of the Combined Cycle Power Station from the work spaces. This is achieved by orientating them towards the West, where a garden provides a rest for the eyes, something that is atypical on such an estate.

LOCATION\Boroa Industrial Estate, Amorebieta – Etxano, Spain

ARCHITECT
Javier Pérez Uribarri, Xabier Aparicio Ortega
FIRM
ACXT Architects
DESIGN
Jose Cavallero, Nuno Lobo, Oscar Malo, Roberto Fdez de Gamboa.
DETAILED DESIGN
Roberto Aparicio, Cristina Lamikiz, Oscar Costa, Marc Rips, Jose Cavallero, Nuno Lobo, Fernado Ortega, Marina Duran, Oscar Malo, Rafael Pérez, Arrate Achalandobaso, Ziortza Bardeci, Francisco García, Angel Gómez, Amaia Lastra, Ibai Ormaza, Igor Udaondo
AWARDS
CSCAE 2009, 8th Architecture Biennial Sao Paulo. Selected work for the exposition
AREA
19006 m²
PHOTOGRAPHER
Aitor Ortiz

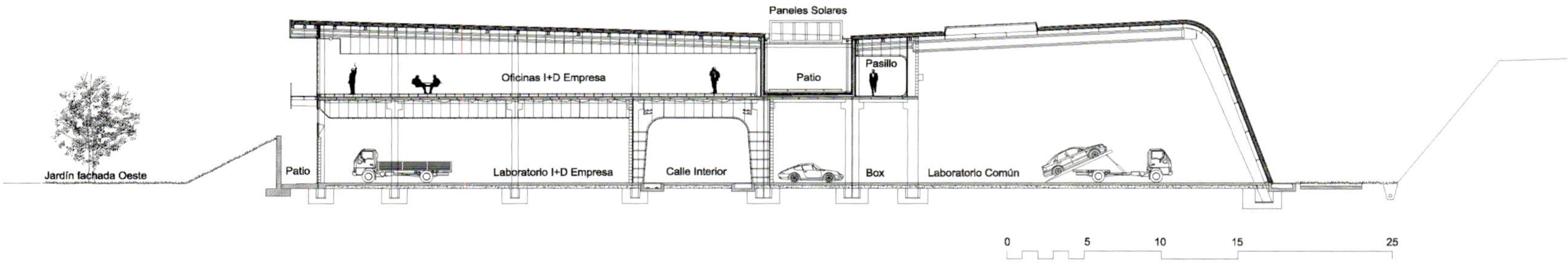

受原型车和赛车车身形态的启发，AIC 汽车智能中心的主建筑群里的所有建筑之上都罩着一个巨大的流线形银色屋顶，使之构成一个统一的整体。主建筑主要由两个部分构成：位于主厅四周的社交区以及研发项目区。

研发项目区各种设施的功能布局可以理解成是根据该中心活动的保密性的需要而设置。大楼外，从西向东是一片花园，花园沿着外墙与周围地形完全融合，呈现出美丽的庭院景观，形成了中心人口的“防护板”。“防护板”充当着屏风的作用，这样各个公司的实验室里安装的小型机器就不会给人们造成视觉冲击。

大楼的第一层是细长的用于内部通行的“街道”，供处于“街道”两边的公司共用。覆盖在“街道”之上的楼面上的通道起着同样的通道作用，供进入办公室的人们使用。位于建筑最高位置的封闭区域设有公用实验室。流线形屋顶的下端是为未来共享项目预留的。在屋顶的东端，东立面就像一道从屋顶翻折到地面的褶皱，将联合循环电厂从观众的视线中阻挡出去。覆盖的“街道”使建筑物突出表达了中心想要表达的团结形象。同时还有很多其他好处，例如，这种设计使整个中心免受 Vizcaya 气候的冲击，同时也对凌乱的空间加以控制，使之显得整齐。

建筑最前端设计的是社交区域。朝南的大厅标志着“街道”已经到了末端。社交区域各职能部门散落在大厅的周围。大厅各个方向安装的巨型玻璃门使大厅与各个部分相通。其中培训区各单位就整合在一个铝制方形空间中，楼下是会堂，楼上是供培训用的教室。

中心的行政管理部门位于屋顶最具动感的部分；这一部分受赛车空气吸入口的启发而设计成褶皱形态。这种设计使进入管理部门的门廊显得更加大气壮观，同时也使管理人员站在办公室就能将四周景观一览无余。

与城市的融合

AIC 汽车智能中心的建筑群位于远离市中心的工业园区，其设计策略一方面是使人们在经过工业园区的高速公路上一眼就能注意到该中心；另一方面也要让在中心工作区工作的人们不会只看到联合循环电厂。为了达到这种效果，该设计将整个楼群朝西，让人们从楼群里望出去满眼都是花园景观，这在工业园区的建筑中是不多见的。

PLANTA CUBIERTA

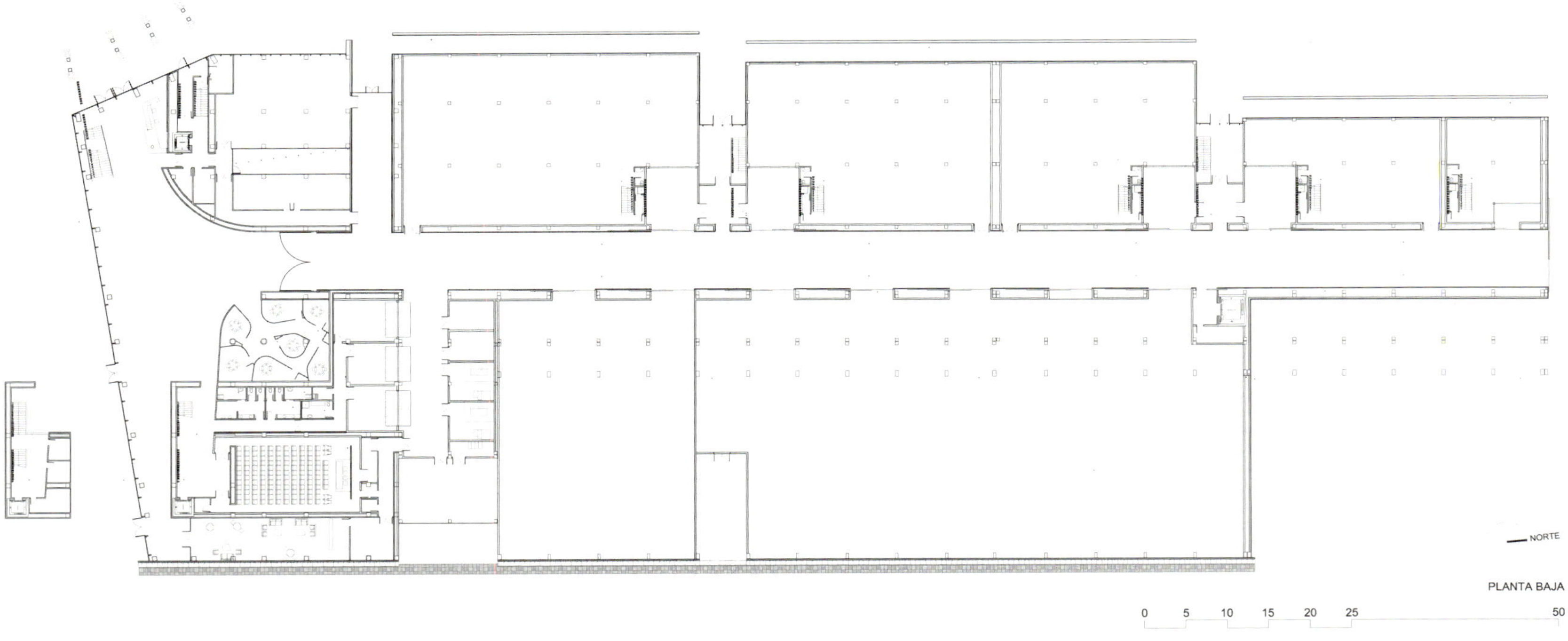
NORTE
PLANTA BAJA
0
5
10
15
20
25
50

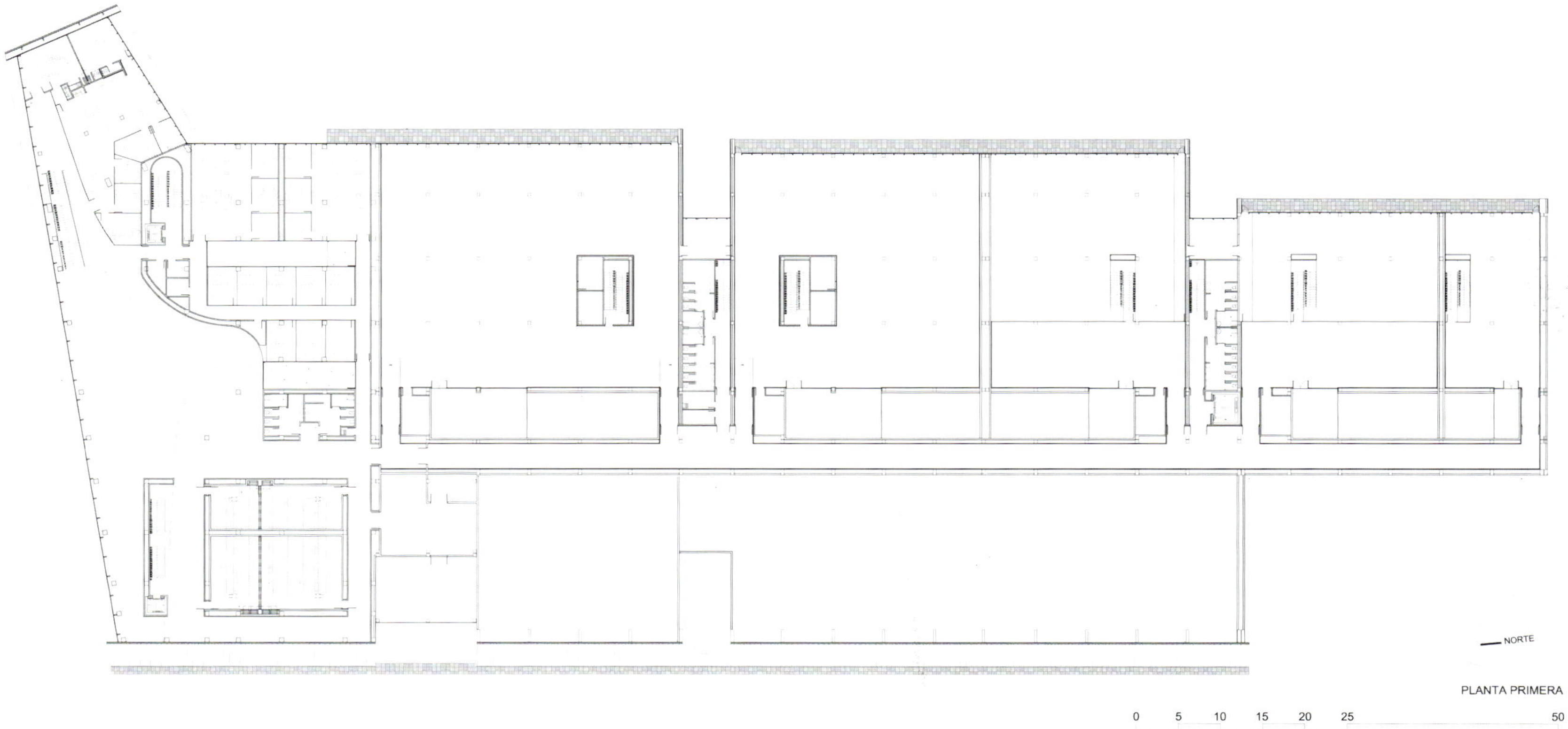
NORTE
PLANTA PRIMERA
0 5 10 15 20 25 50

RESERVED

555 Mission Street

555 Mission Street, the first LEED-CS Gold high-rise office tower in San Francisco, is a thirty-three-storey office building incorporating stunning views of San Francisco City and Bay. The façade, made entirely of floor-to-ceiling glass panels accented with glass and metal fins, creates a textural layer that beautifully captures and reflects light. The unique curtain wall system with glass fins adds a rich texture to the overall expression. The building is at the center of the Mission Street transit corridor, which is within easy walking distance of the city's prominent restaurants, hotels, and the Union Square shopping area, as well as adjacent to the Transbay Terminal, the hub to many of the areas transit lines.

The project was instrumental in enhancing the evolving downtown mid-block open space pattern to maximize light, air, and open public space opportunities. The form of 555 Mission responds to the neighboring context and the Mission Street transit corridor, emphasizing verticality and slenderness. A major outdoor plaza was created in the space between 555 Mission and the 101 Second Street, forming a clear connection between the two towers and reinforcing the key north/south mid-block pedestrian link beginning at Market Street.

Sustainable features that contributed to the building's LEED-CS Gold rating include elimination of refrigerants in all building heating and cooling systems that emit compounds that contribute to ozone depletion, recycled content from the building structural steel and in the aluminum curtain wall, a building mechanical design that brings in 30% more fresh air to occupied spaces than required by code, low VOC paints and carpet systems in the building that provide better indoor air quality, water conserving fittings and features, and a highly reflective and lightly colored "cool roof" system that reduces the effect of urban heat islands.

ARCHITECT
Heller Manus Architects, in collaboration with Kohn Pedersen Fox Associates (KPF)

SUSTAINABLE FEATURE
LEED-CS Gold (First LEED Gold office tower in San Francisco)

AREA
58064 m²

PHOTOGRAPHER
Heller Manus Architects, Tishman Speyer Properties, Russell Abraham

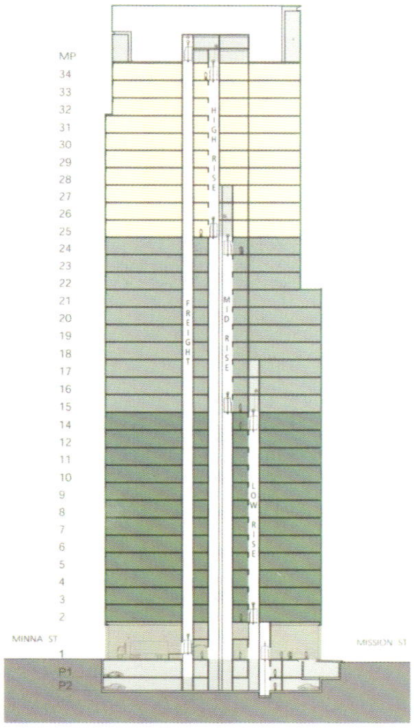
MP
34
33
32
31
30
29
28
27
26
25
24
23
22
21
20
19
18
17
16
15
14
12
11
10
9
8
7
6
5
4
3
2
1
P1
P2
MINNA ST
MISSION ST
HIGH RISE
MID RISE
LOW RISE
FREIGHT

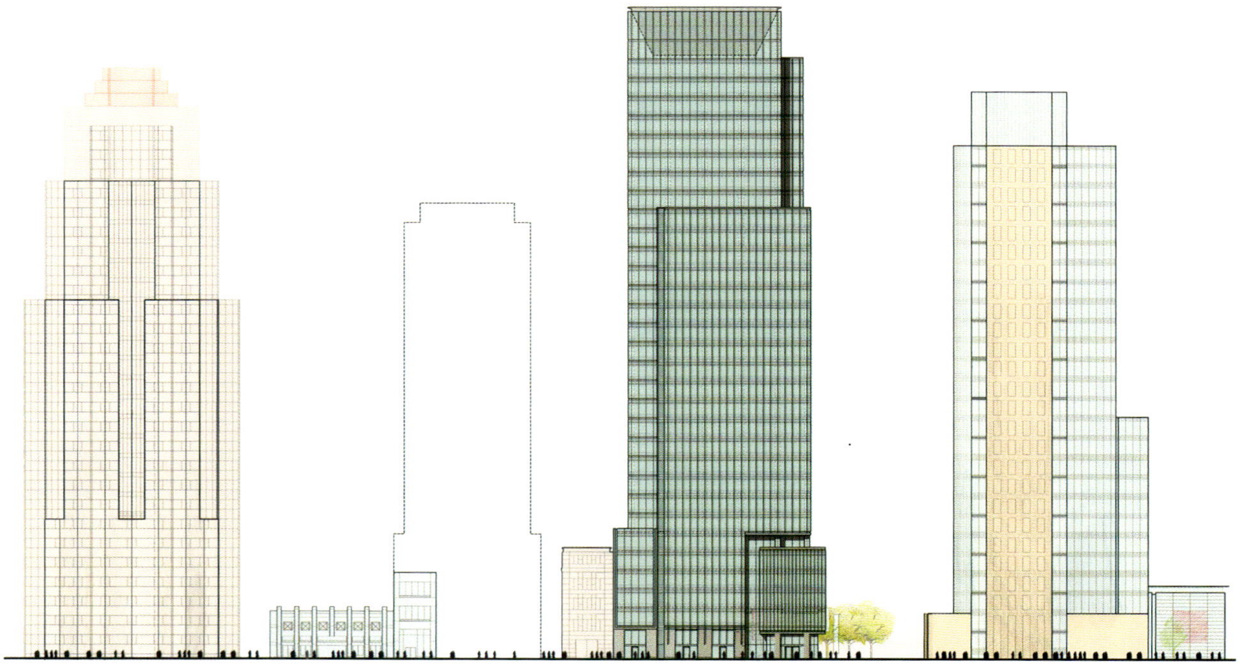

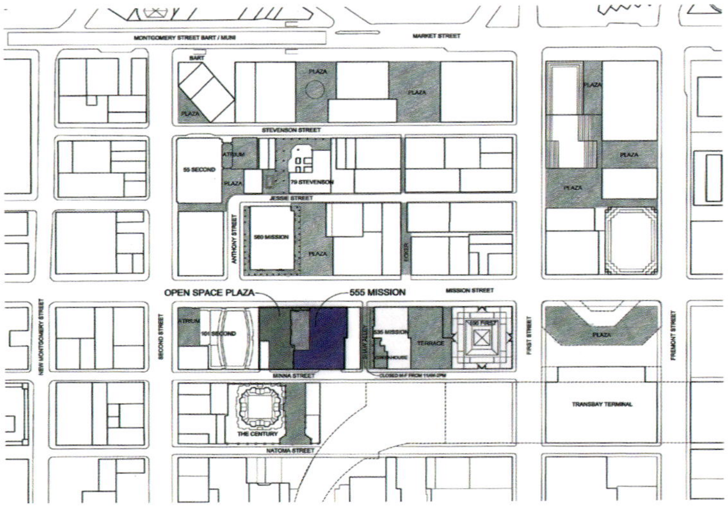
MARKET STREET
STEVENSON STREET
JESSIE STREET
OPEN SPACE PLAZA
555 MISSION
MISSION STREET
MINNA STREET
NATOMA STREET
TRANSBAY TERMINAL

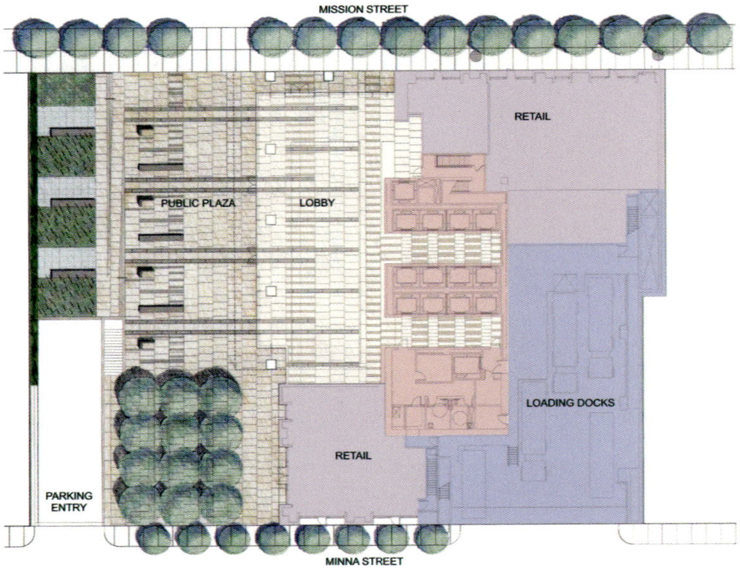
MISSION STREET
RETAIL
PUBLIC PLAZA
LOBBY
LOADING DOCKS
RETAIL
PARKING
ENTRY
MINNA STREET

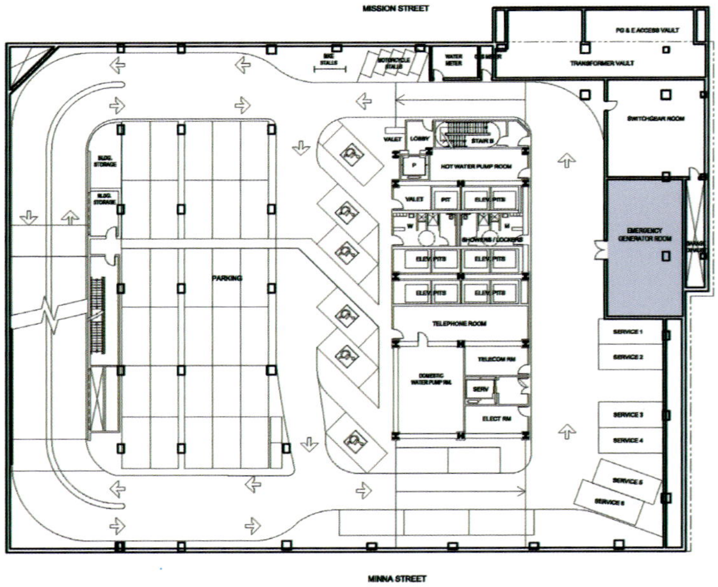
MISSION STREET
TRANSFORMER VAULT
SWITCHGEAR ROOM
HOT WATER PUMP ROOM
PARKING
TELEPHONE ROOM
MINNA STREET

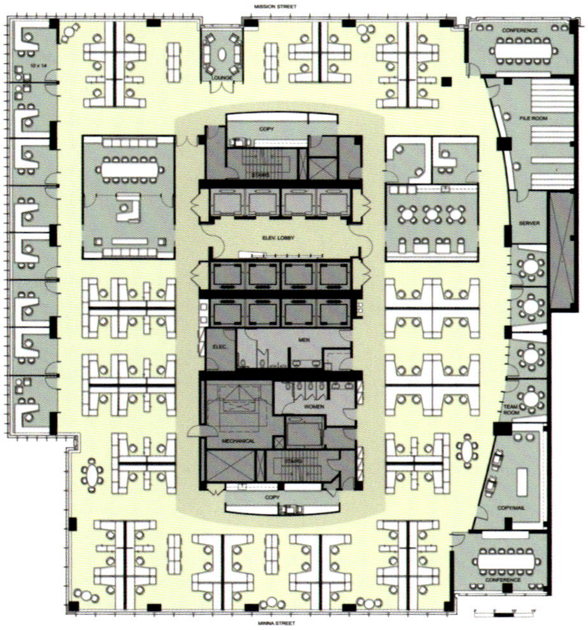

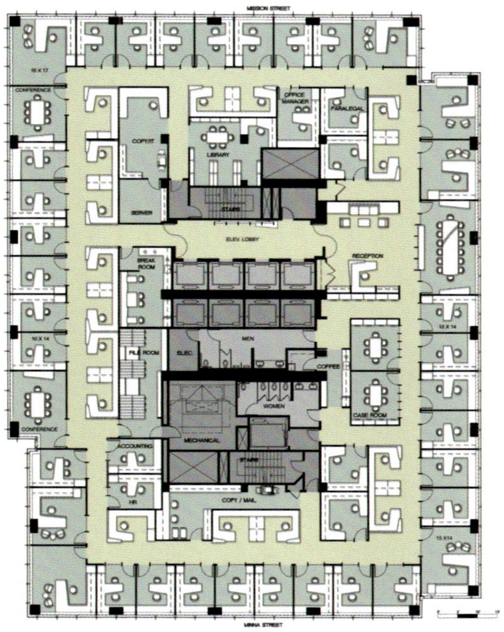

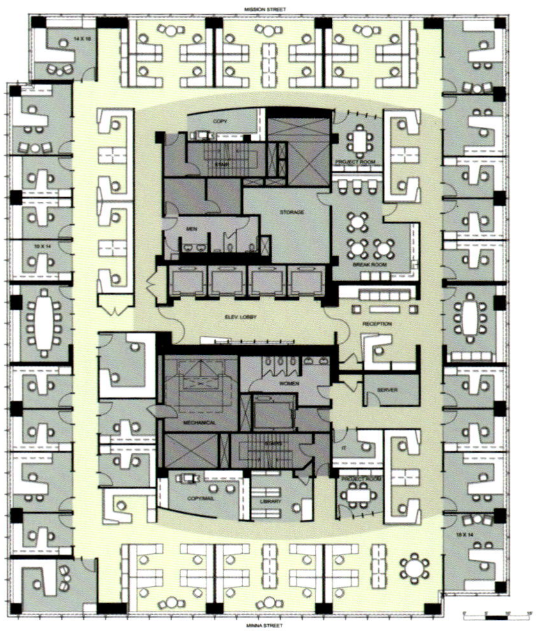

大使街 555 号大楼是旧金山市首个获得 LEED 绿色建筑认证的多层多用型建筑，这座 33 层高的写字楼将旧金山市和海湾的迷人景色尽收眼底。整个大楼立面由从地板到天花板的玻璃板覆盖并用玻璃和金属翼片加固，形成独特的层次，并通过对光的吸收和反射，形成完美的视觉效果。同时，玻璃翼片的使用使整座建筑的层次更加丰富，增强了整体的表现效果。该建筑位于大使街走廊的正中心，步行能很快在附近找到市区最好的餐厅、酒店和联合广场购物区，并靠近通达各地区的交通枢纽中心——旧金山著名的港湾枢纽区。

该设计有助于将市中心的中间街区逐步变成开放式的空间布局，最大限度地拥有采光、空气流通和开放公共空间。大使街 555 号大楼的设计充分考虑了邻近的建筑风格和背景，其特别强调的垂直感和狭长感与周围景观交相辉映；该大楼的建成，使得该大楼和第二大道的 101 号大楼之间的空地上自然形成了一个巨大的户外广场。广场将两座高楼连接起来，使商业街上南北方向最繁华的中街地段的行人流动更加畅通。

该大楼具有显著的可持续性发展的特色，包括：在整幢大楼的制热和制冷系统中不使用排放破坏臭氧层的化合物的制冷剂、建筑结构钢架和铝制幕墙系统中使用可再生性材质、建筑结构使单位空间里的新鲜空气流通率比规定的标准高出 30%、较少使用挥发性油漆和地毯装置以提高室内空气质量、各具特色的储水装置、降低城市热岛效应的反光性能良好的浅色“凉爽型屋顶”系统等等，这些特色使得该大楼获得 LEED 绿色建筑认证。

CSET-Center for Sustainable Energy Technologies

ARCHITECT
Mario Cucinella, Elizabeth Francis, Angelo Agostini, David Hirsch

IN COLLABORATION WITH
Eva Cantwell, Richard Ceccanti (3D), Francesco Fulvi,
Caterina Maciocco(STRL), Giuseppe Perrone, Natalino Roveri (models),
Luca Stramigioli, Debora Venturi (Environmental Strategies)

LOCAL ARCHITECT
NADRG, Ningbo - P.R. of China, Guo XiaoHui

AREA
1200 m²

PHOTOGRAPHER
Daniele Domenicali

The new building will provide laboratory, office and seminar accommodation and has been designed to serve as an exemplar building, demonstrating state-of-the-art techniques for environmentally responsible, sustainable construction and energy efficient internal environmental control. It has been designed to minimize its environmental impact by promoting energy efficiency, generating its own energy from renewable sources, and using locally available materials with low embodied energy wherever possible.

The main function of the building is to provide a specialist research laboratory for staff and postgraduate students within the new Center for Sustainable Energy Technologies. The focus of research will be new and renewable energy systems and components for both domestic (housing) and nondomestic (commercial and public) buildings. The building will provide a facility for developing and testing the new technologies in collaboration with local and international industrial partners.

Linked to the laboratory facilities, will be a workshop for fabrication of experimental rigs etc., and the development of new components. The tower incorporates a research studio/teaching room, a resource room, as well as offices, meeting rooms and permanent display space. The exhibition space will provide a platform for communicating the latest developments in sustainable energy and construction technologies, both regionally in China and internationally.

Institute of Sustainable Development Low Carbon Design – Environmental Design Strategy

The CSET building has been designed to respond to diurnal and seasonal variations in ambient conditions by means of a five-point environmental design strategy: High Performance Envelope, Exposed Thermal Mass, Daylight & Solar Control, Natural Ventilation to Tower, and Piped Ventilation to Laboratory & Workshop.

In this way, the building is designed to minimize the need for additional energy for heating, cooling and ventilation.

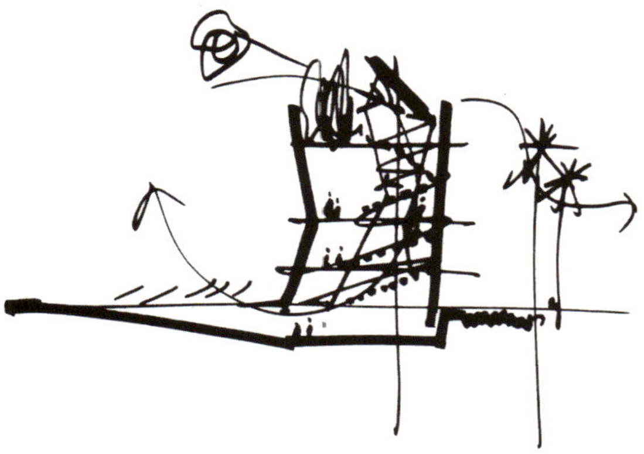

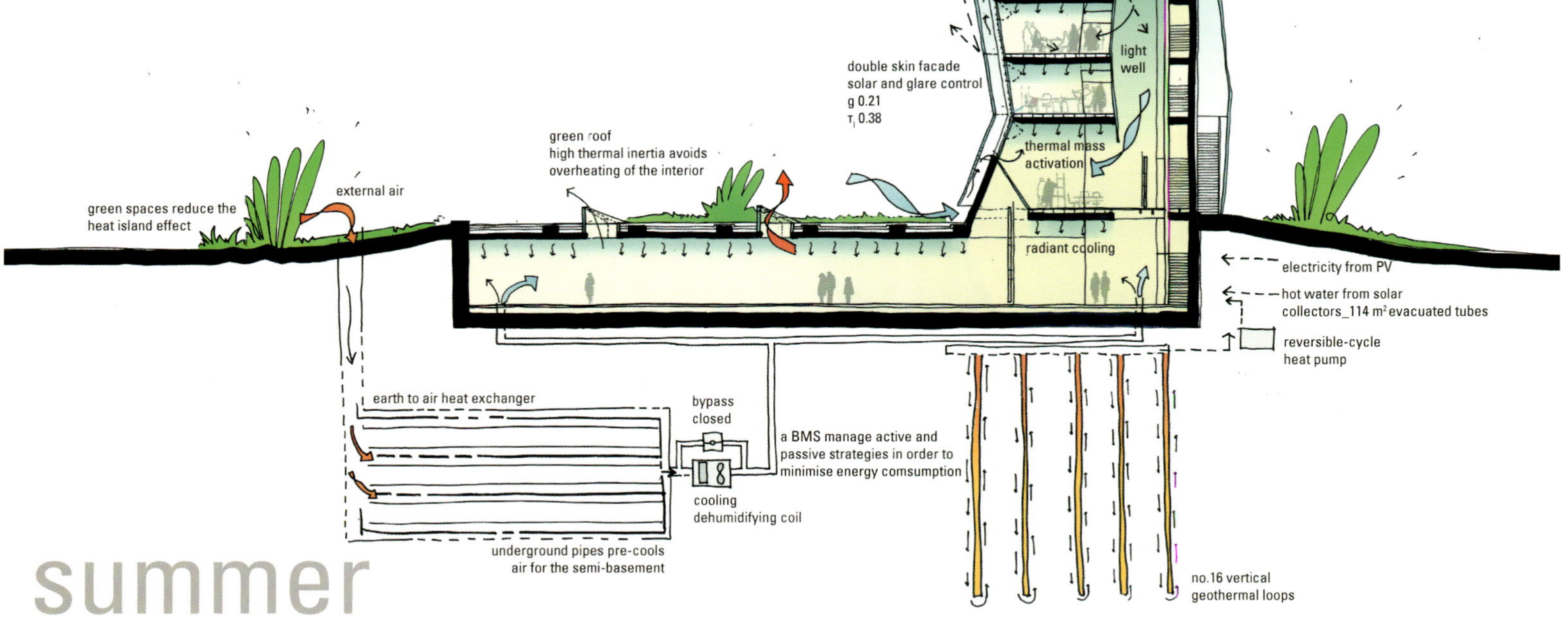

21 june_12 am
solar cooling
a chiller, powered by hot water
from solar tubes, pre cools external
air for ventilation of the tower
exhaust
air
closed
high thermal inertia of the
exposed concrete surfaces
light
well
double skin facade
solar and glare control
g 0.21
τ_i 0.38
green roof
high thermal inertia avoids
overheating of the interior
thermal mass
activation
green spaces reduce the
heat island effect
external air
radiant cooling
electricity from PV
hot water from solar
collectors_114 m² evacuated tubes
reversible-cycle
heat pump
earth to air heat exchanger
bypass
closed
a BMS manage active and
passive strategies in order to
minimise energy comsumption
cooling
dehumidifying coil
underground pipes pre-cools
air for the semi-basement
no.16 vertical
geothermal loops
summer

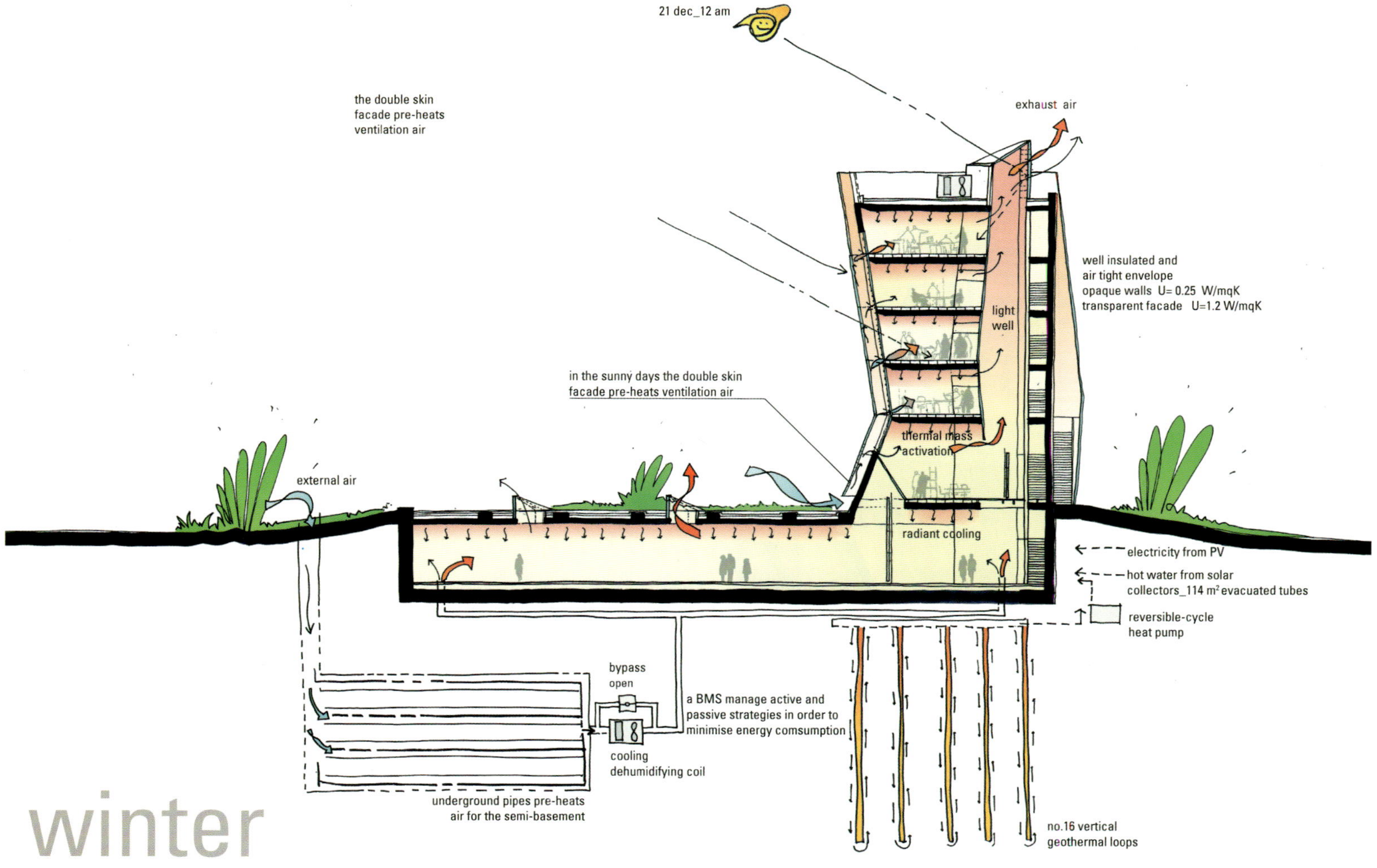

21 dec_12 am
the double skin
facade pre-heats
ventilation air
exhaust air
well insulated and
air tight envelope
opaque walls U= 0.25 W/mqK
transparent facade U=1.2 W/mqK
light
well
in the sunny days the double skin
facade pre-heats ventilation air
thermal mass
activation
external air
radiant cooling
electricity from PV
hot water from solar
collectors_114 m² evacuated tubes
reversible-cycle
heat pump
bypass
open
a BMS manage active and
passive strategies in order to
minimise energy comsumption
cooling
dehumidifying coil
underground pipes pre-heats
air for the semi-basement
no.16 vertical
geothermal loops
winter

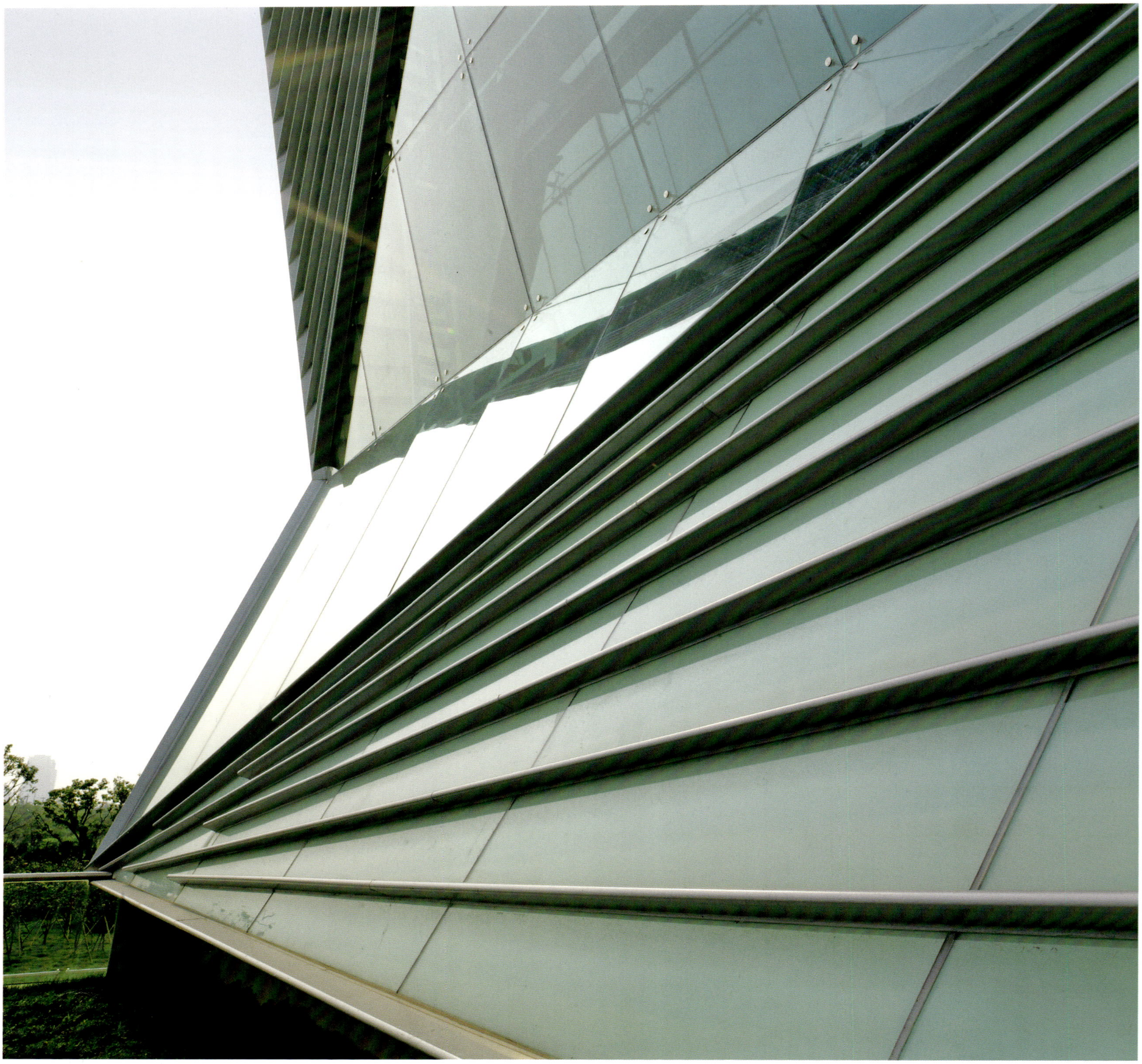

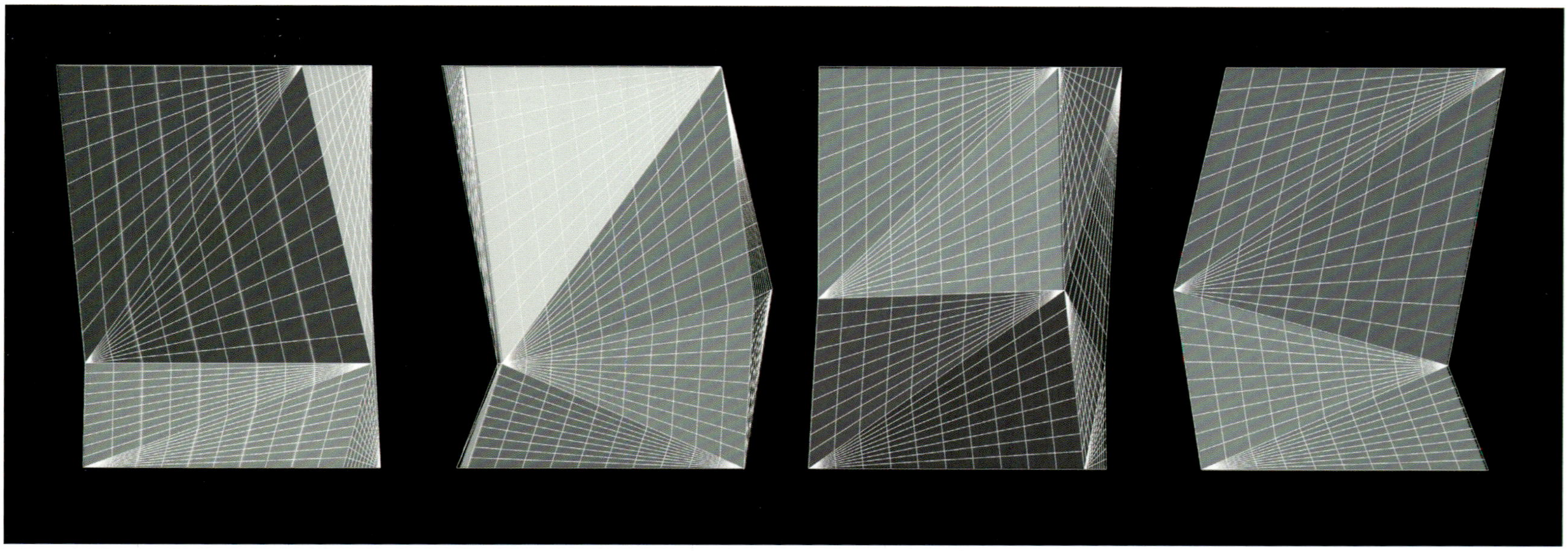

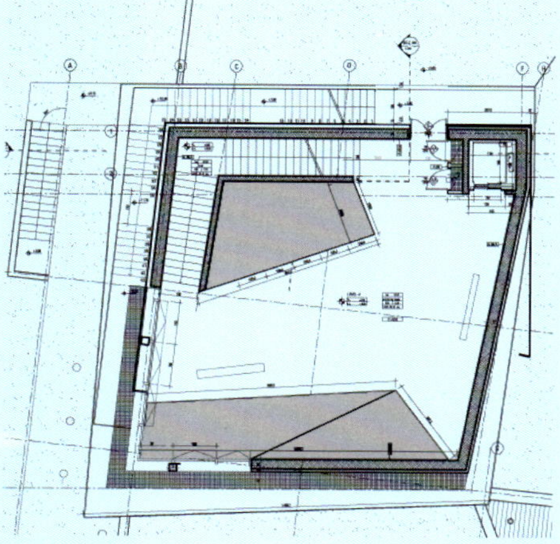

LIVELLO +1

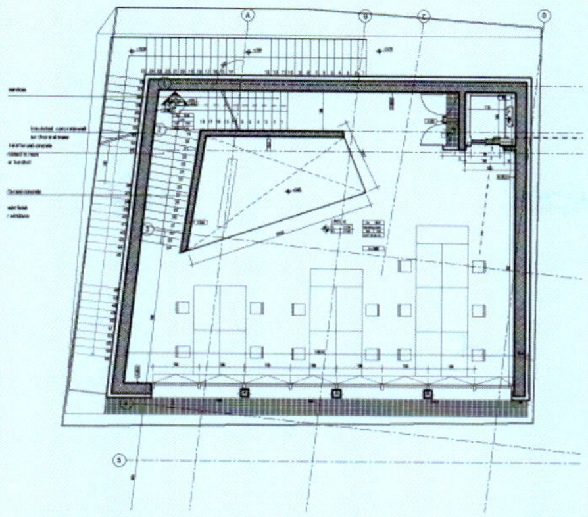

LIVELLO +2

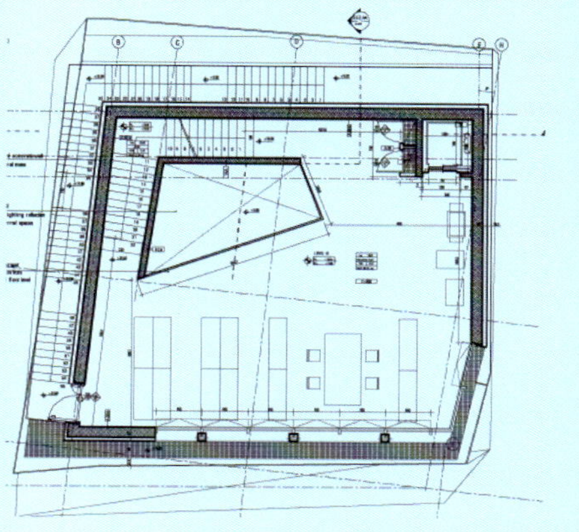

LIVELLO +3

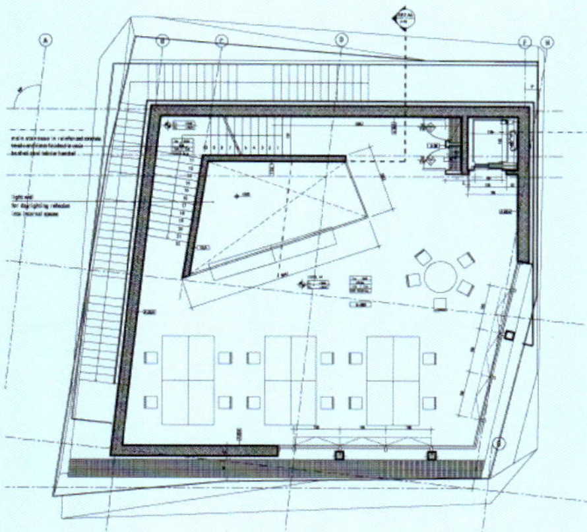

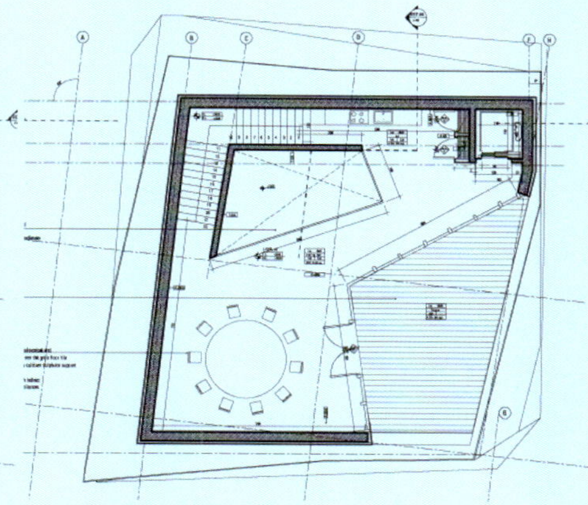

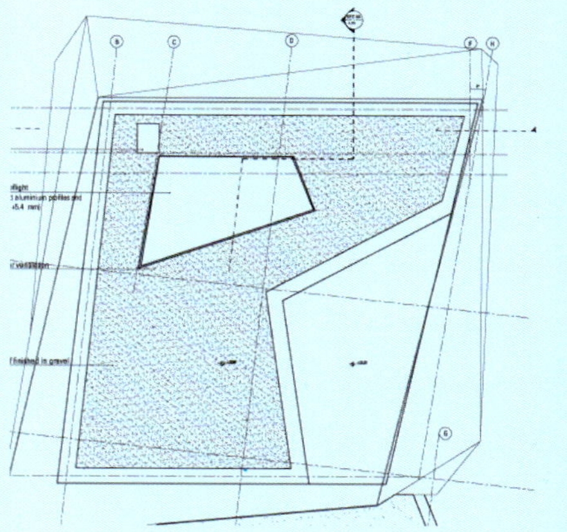

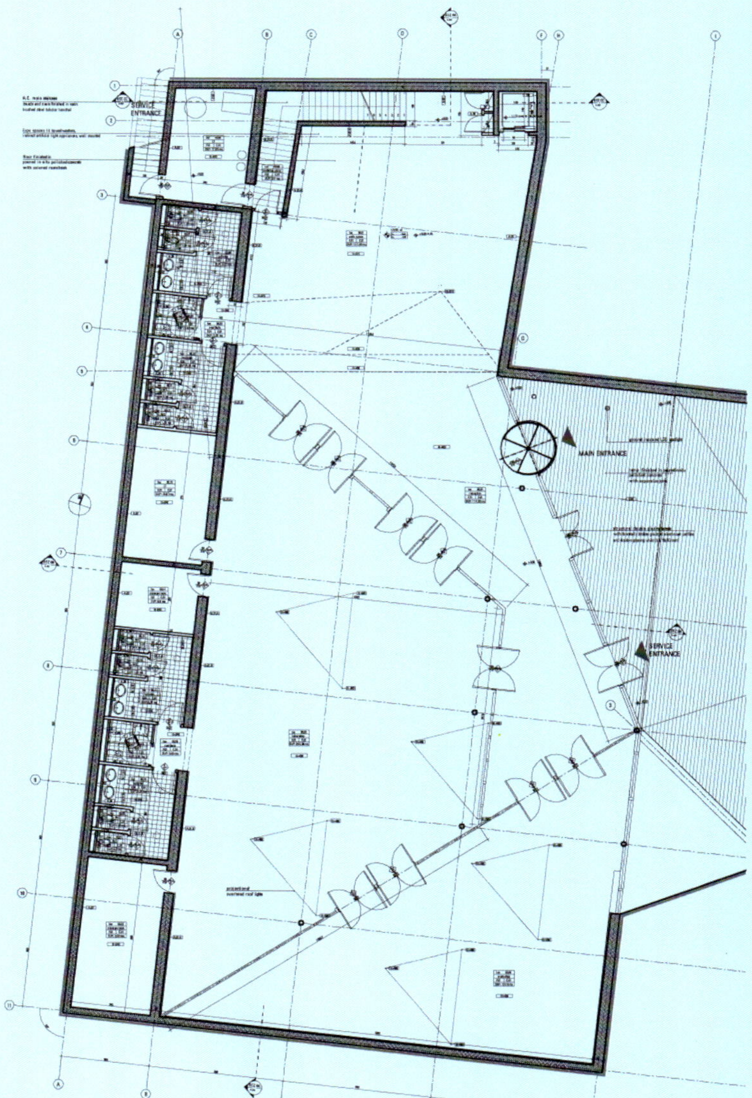

这座建筑通过别具特色的设计，示范了最先进的倡导环保、可持续性和节能理念的建筑技术。通过提高能源利用效率、利用可再生资源实现能源的自给自足、尽量采用当地的低能耗材质等来降低大楼本身对周围环境的影响。该中心既是绿色建筑技术的一个示范载体，也是实验、办公和举办会议的理想场所。

该大楼旨在为可持续能源技术研究中心的研究成员和研究生提供专家级研究实验室，以重点研究用于住宅建筑和非住宅建筑（包括商业和公共建筑）的新型可再生能源系统和建材；还配有用于与当地及国际研究机构合作进行开发和测试新技术的各种先进设备。

中心除了配有实验设备以外，还设有用于制造实验设备的车间和研发新建筑材料的部门。其中塔楼上设有研究工作室／教室、资料室、办公室、会议室以及永久的展示区域。这个展览区域为中国和全世界的其他地区在可持续性能源和可持续性建筑等技术方面的最新开发项目提供一个可供交流的平台。

可持续性发展之低碳设计——环保设计策略机构

本建筑通过五个方面的环境设计策略使得建筑能够随昼夜和季节的变化对周围环境做出动态的回应。这五个方面包括：高性能的围护结构设计、蓄热体设计、采光设计和日光控制、塔楼的自然通风设计以及实验室和加工间的管道通风设计。

通过这五个方面的环境设计策略尽可能地减少对加热、冷却和通风等额外能源的需求。

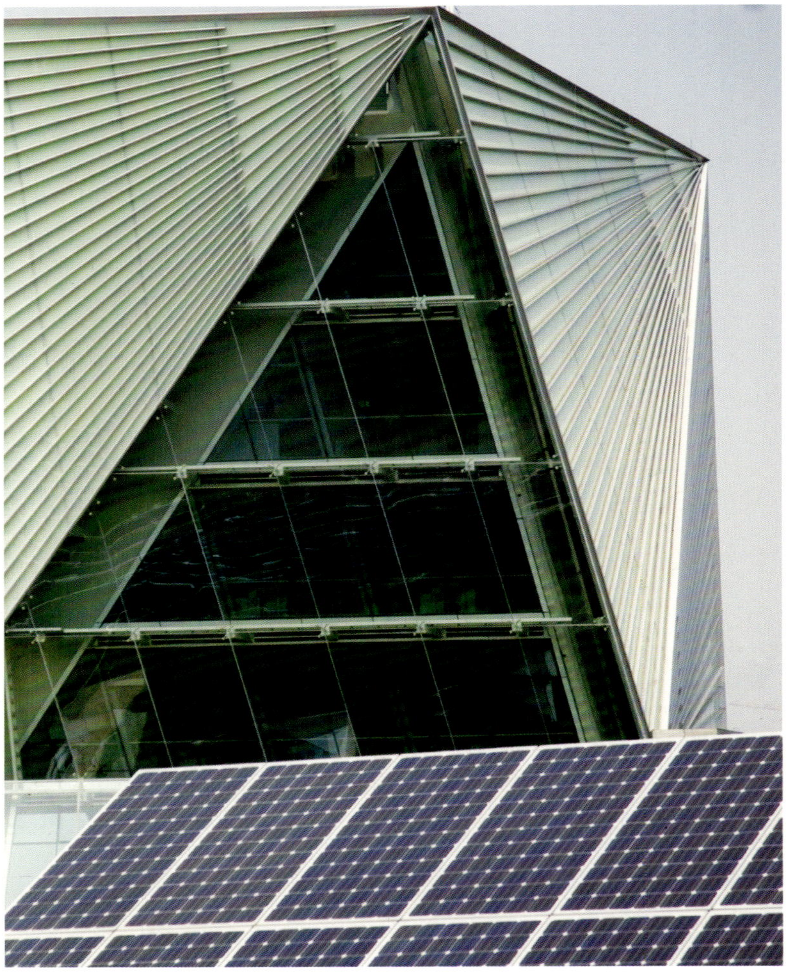

This 411,975m^2 mixed-use project involves new construction of a shopping mall and two offices towers and the facade design and interiors for an existing structure to be fitted out as an office tower. The new buildings total 105,701m^2 and include a 66,183m^2 Iguatemi Shopping Mall, a 22-storey office tower, a 19-storey office tower and underground parking for the development.

The work scope for the 88,630m^2 existing structure includes design of the tower's facades, and interior floor plan layouts and interior design of lobbies and common areas. Santander is the anchor tenant for that office tower. The other 32,000m^2 exist structure (Block B) will be fitted out as office building and theater.

ARCHITECT
Bernardo Fort Brescia, Laurinda Spear
FIRM
Arquitectonica
AREA
105701m^2
PHOTOGRAPHERS
Fancisco Donadio, Courtesy Arquitectonica

LOCATION\São Paulo, Brazil

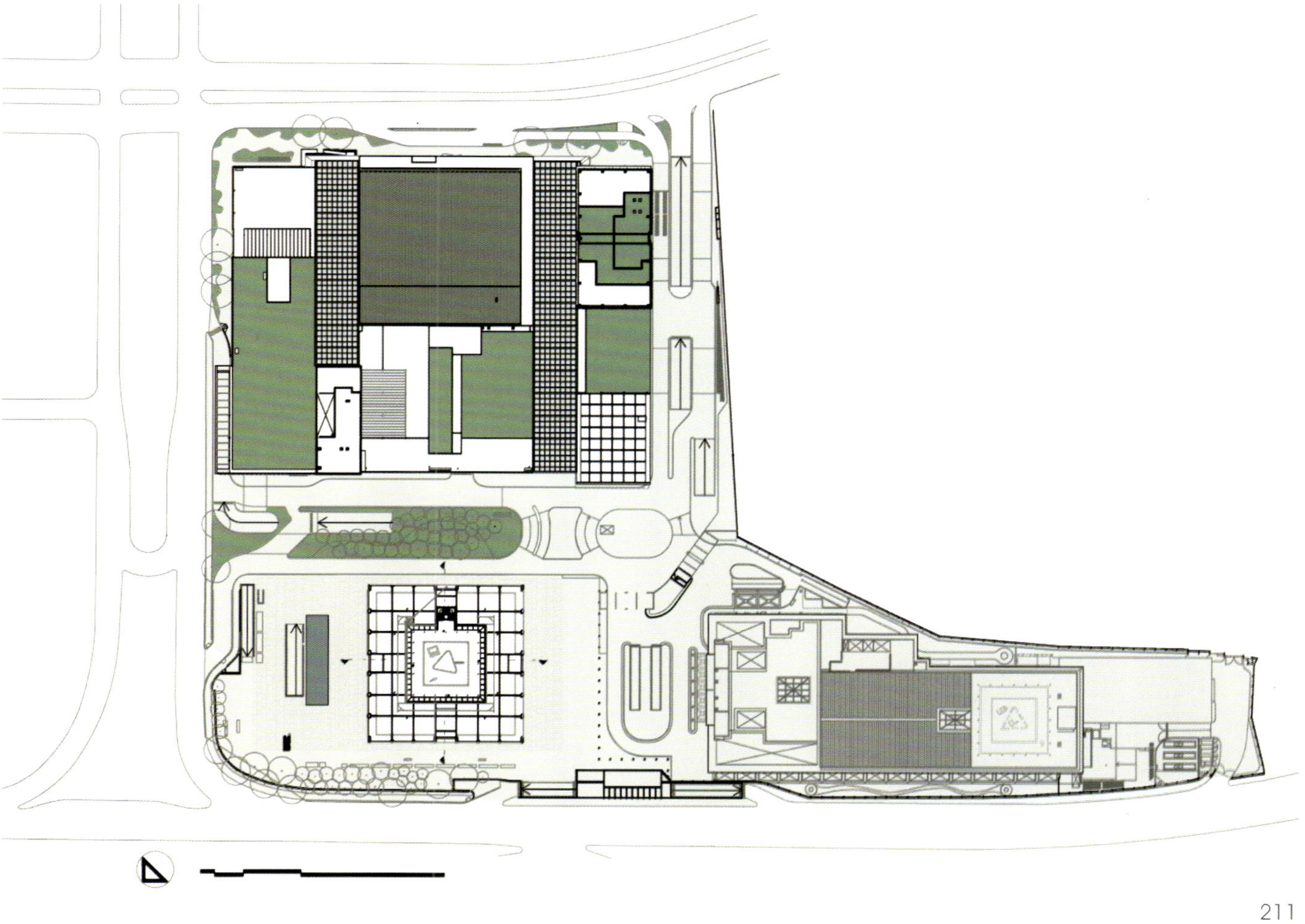

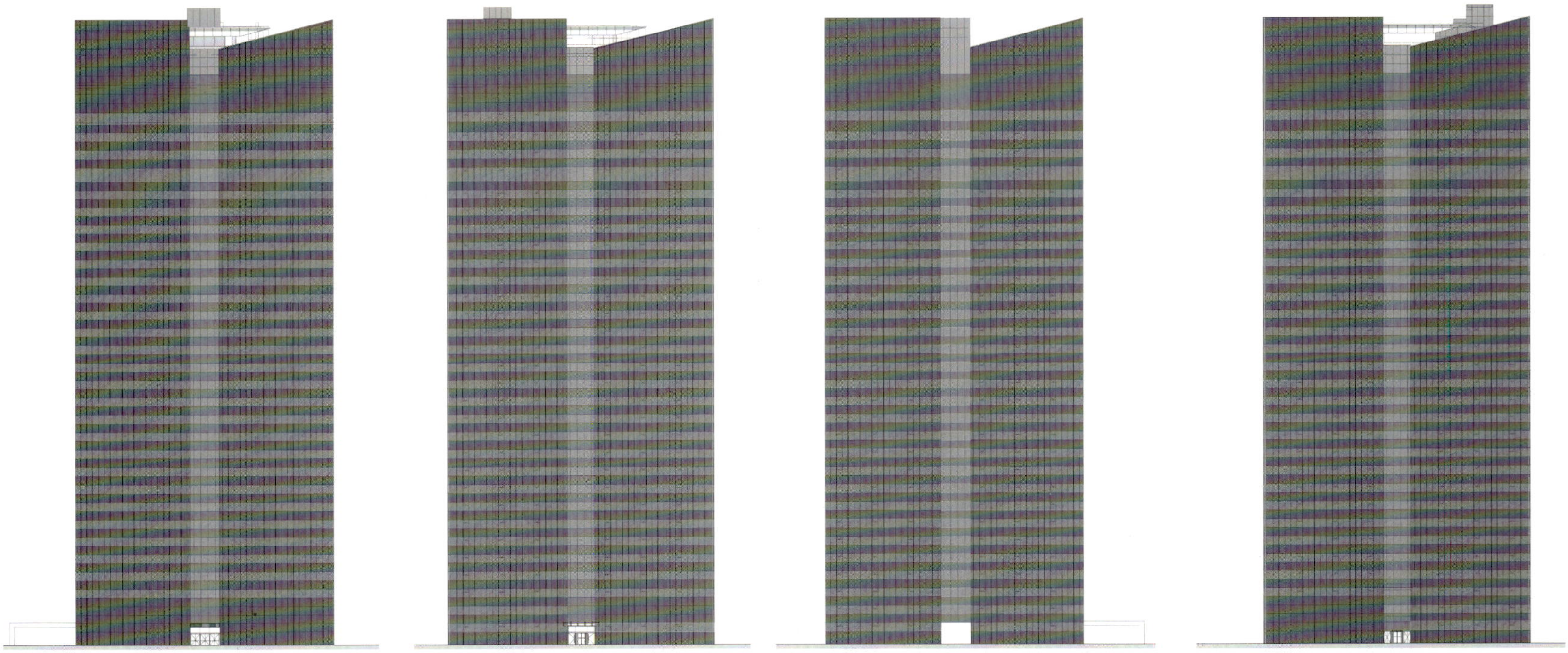

MARG

JK

CHEDI

DASLU

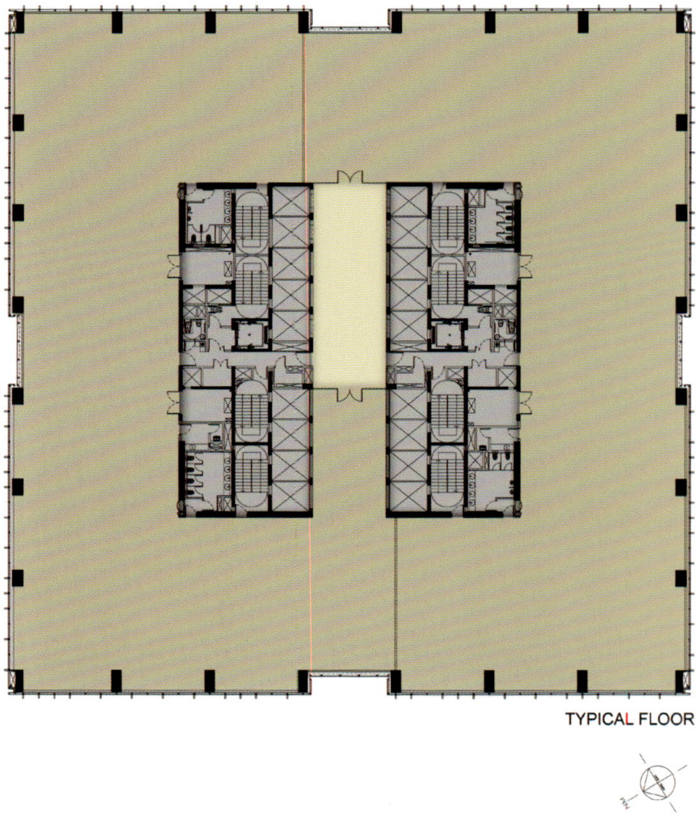

TYPICAL FLOOR

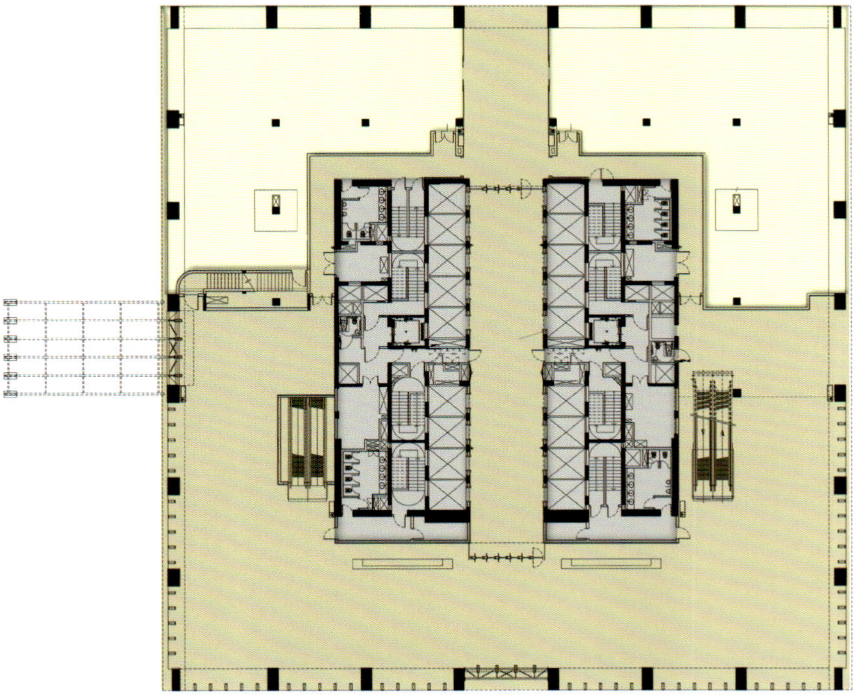

GROUND FLOOR

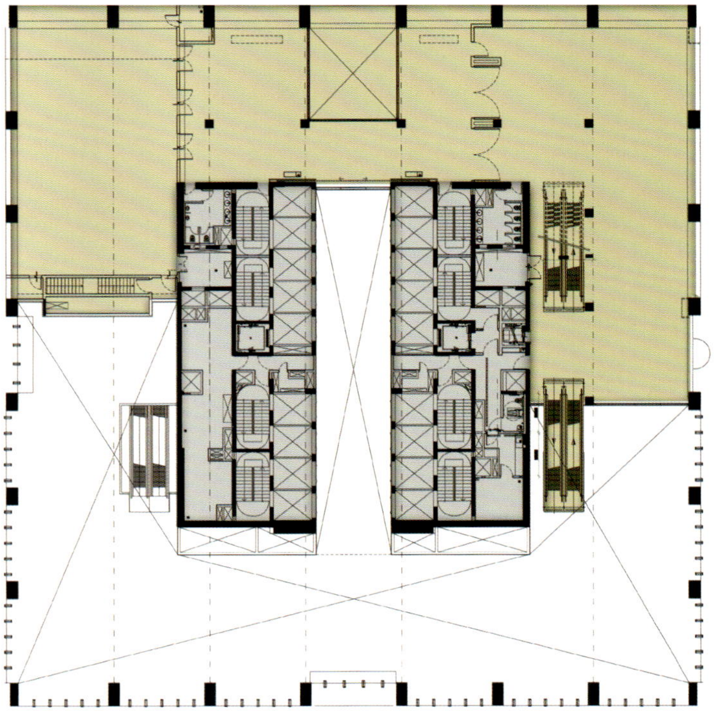

MEZZANINE

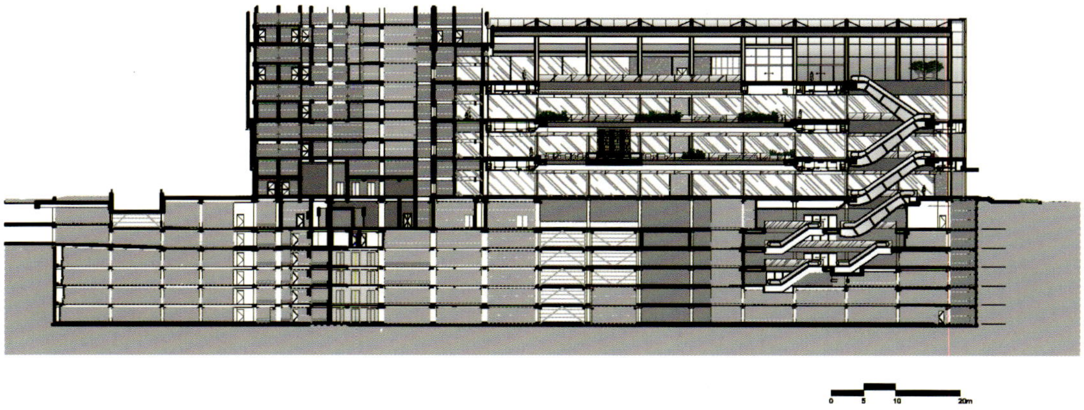
0 5 10 20m

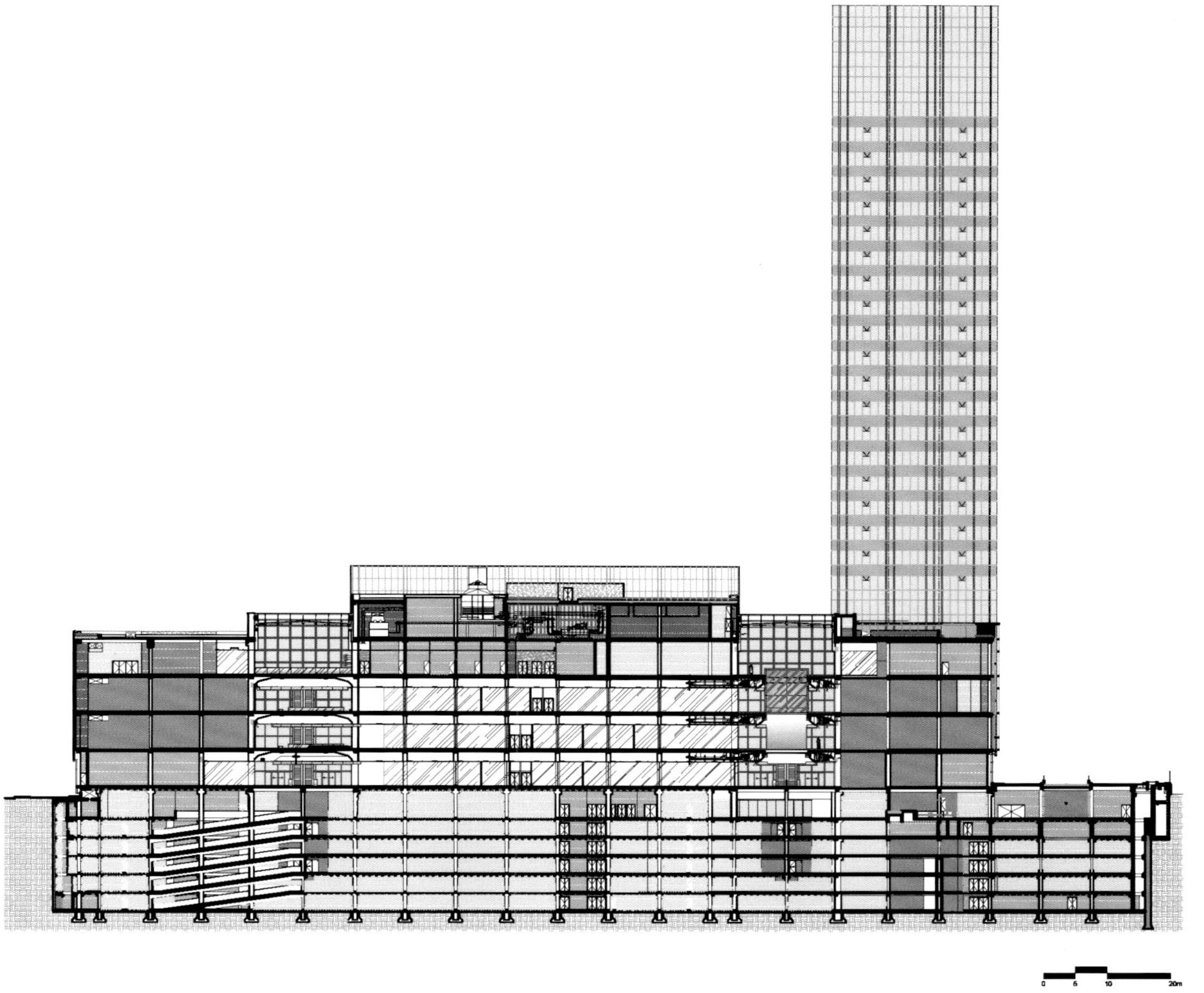

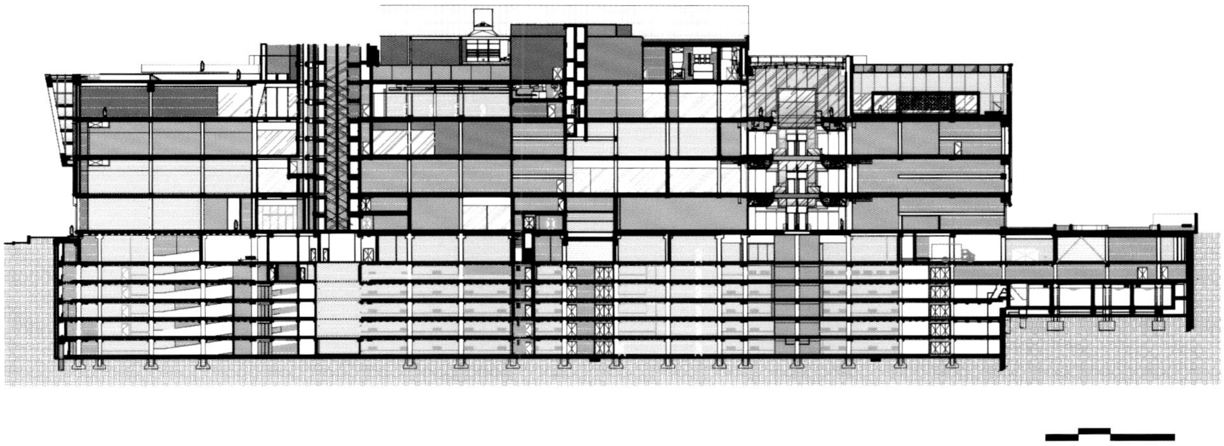

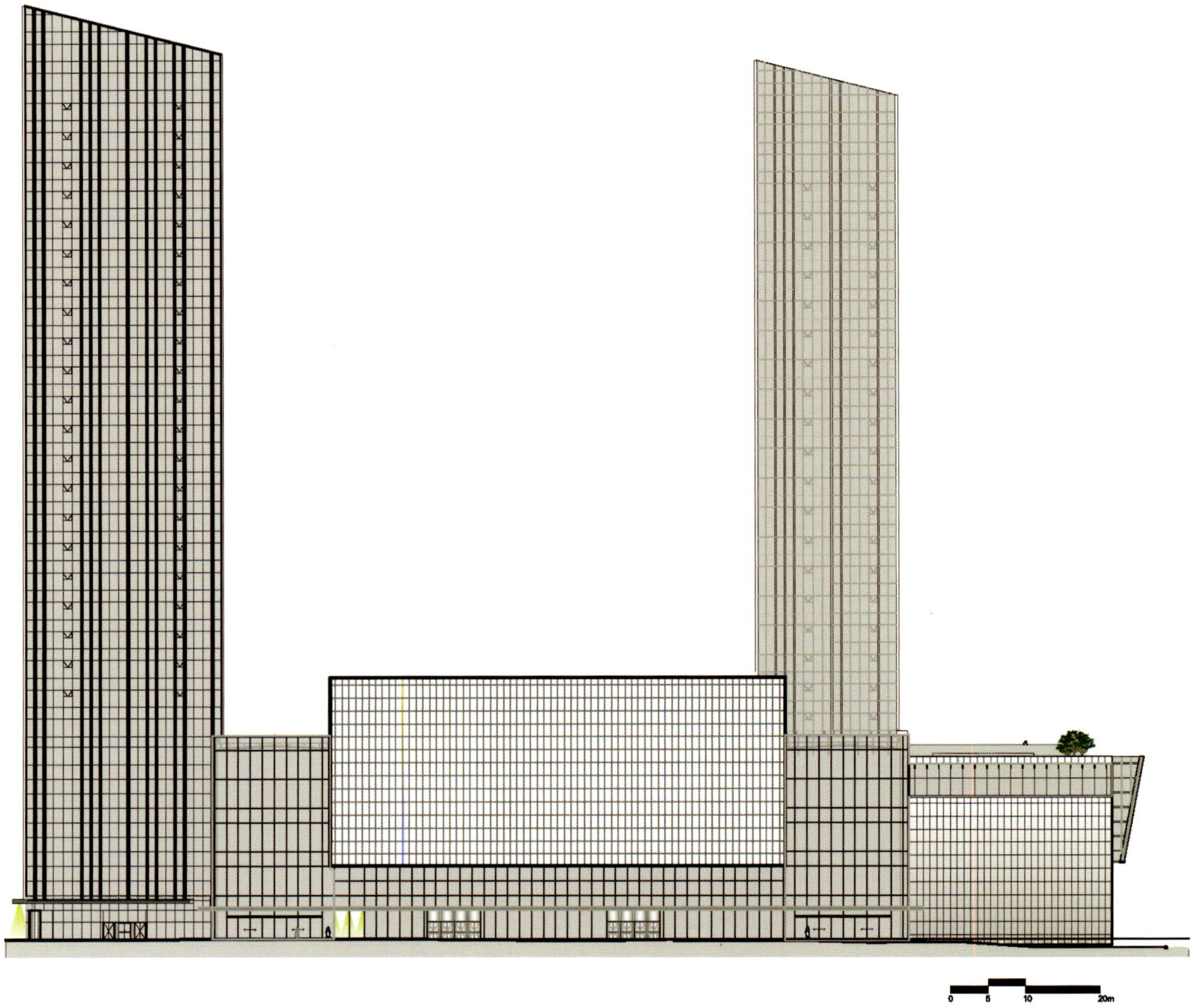
0
5
10
20m

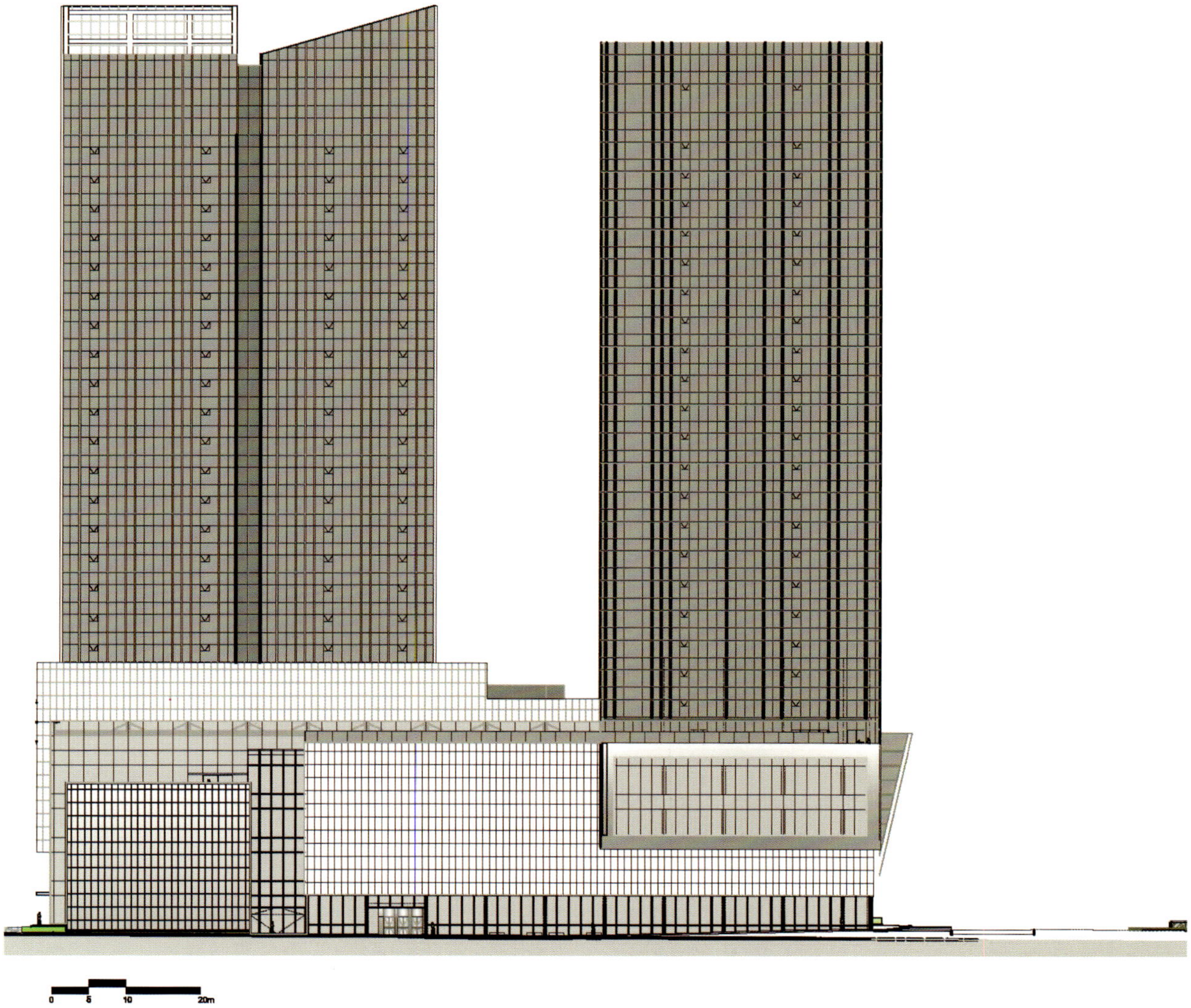
0
5
10
20m

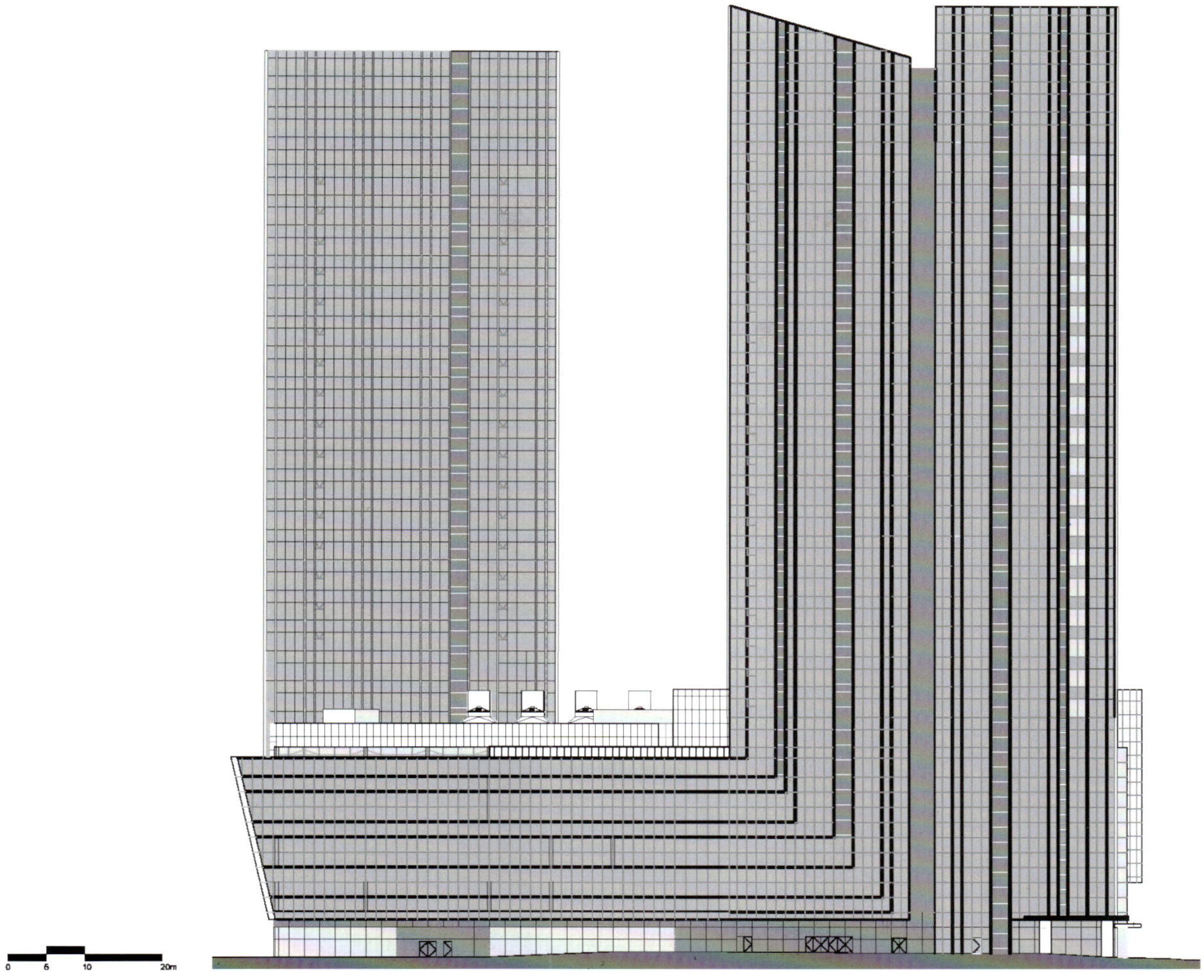
0 5 10 20m

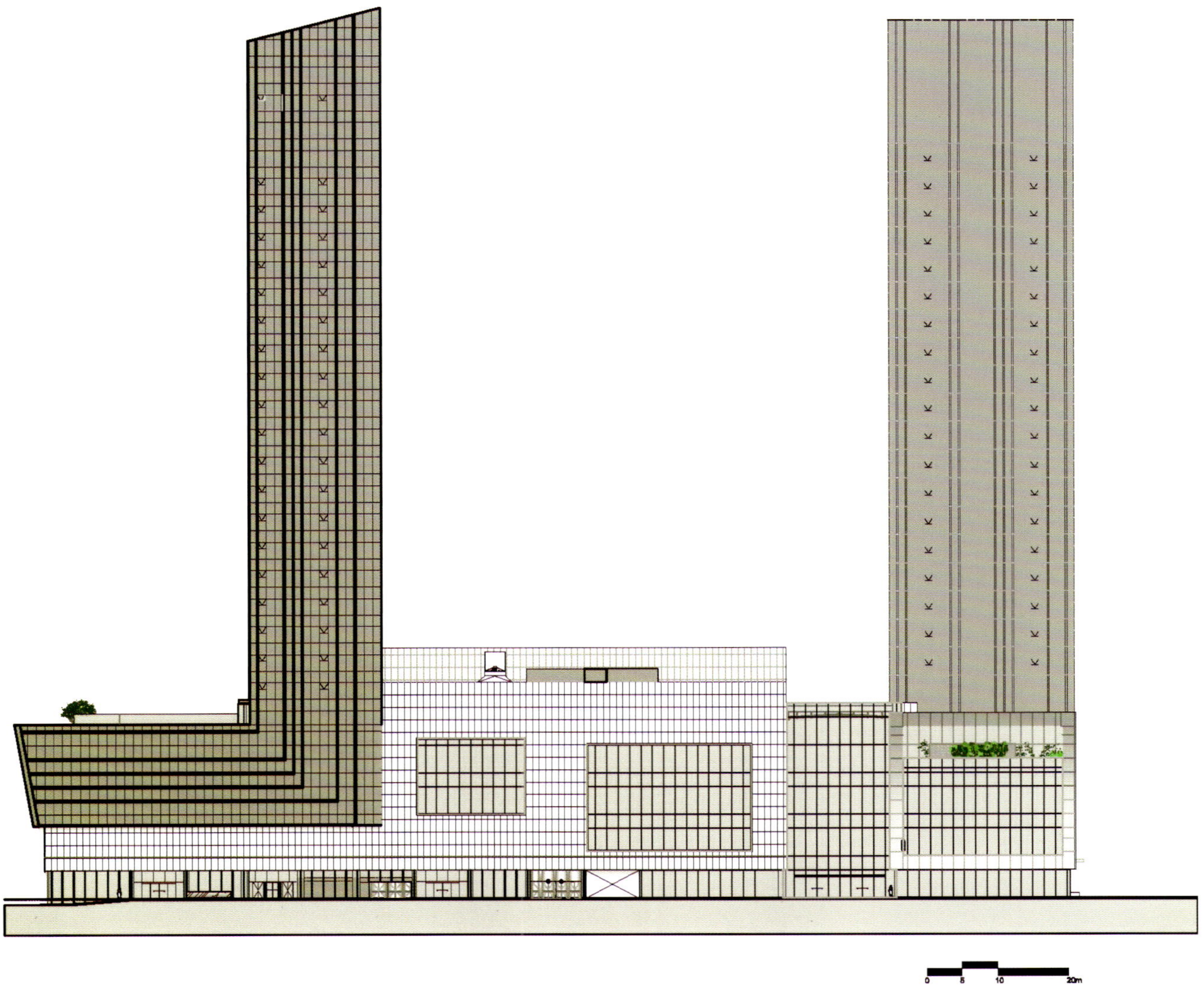
0 5 10 20m

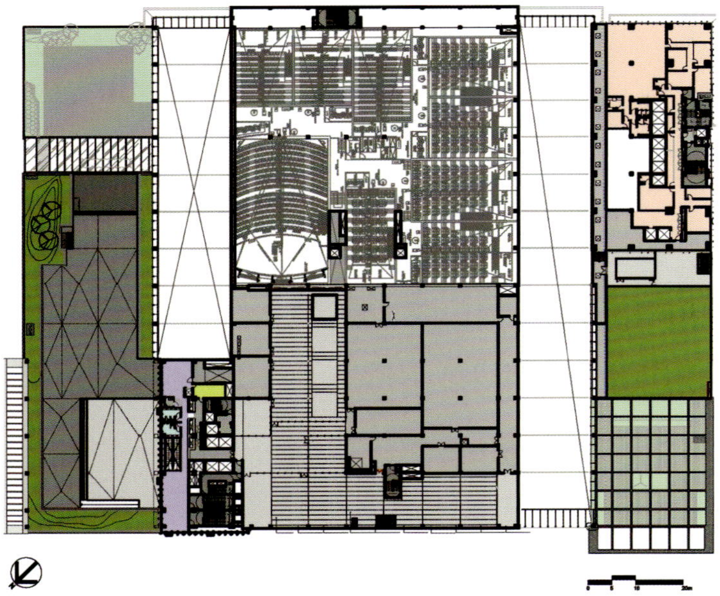

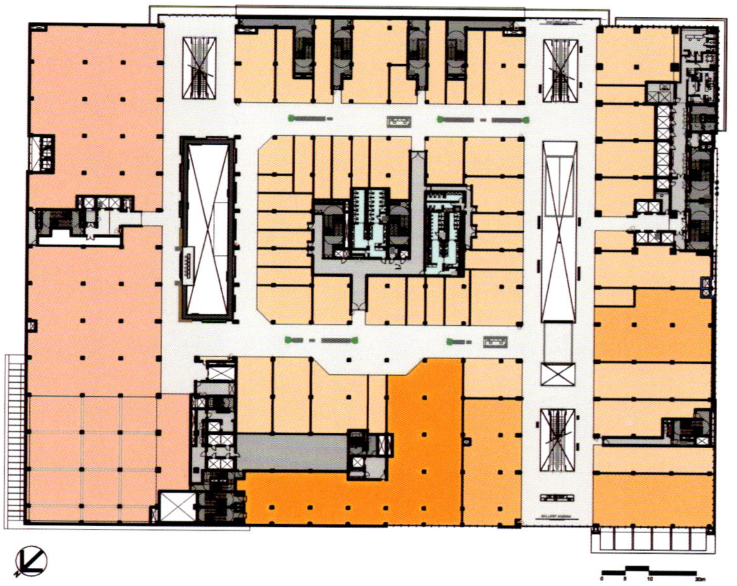

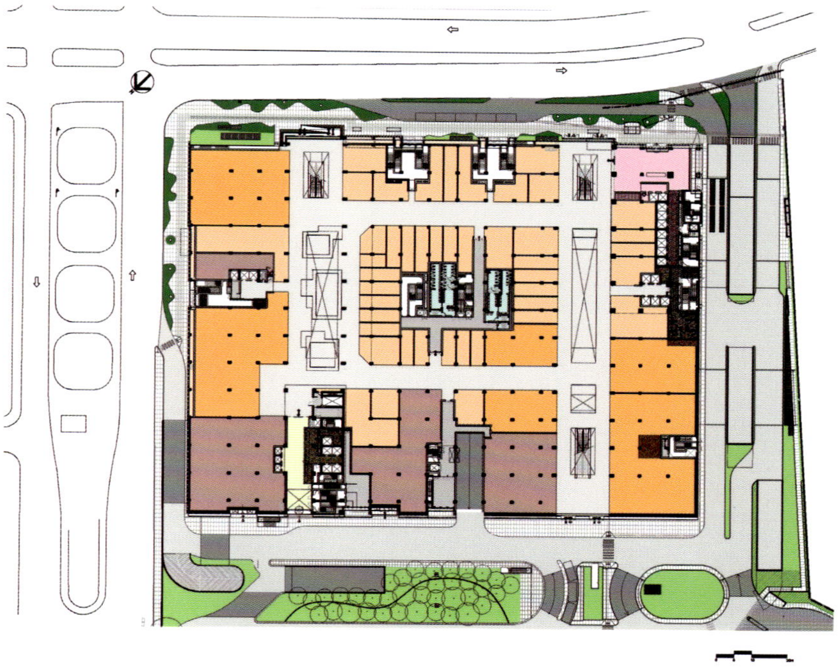

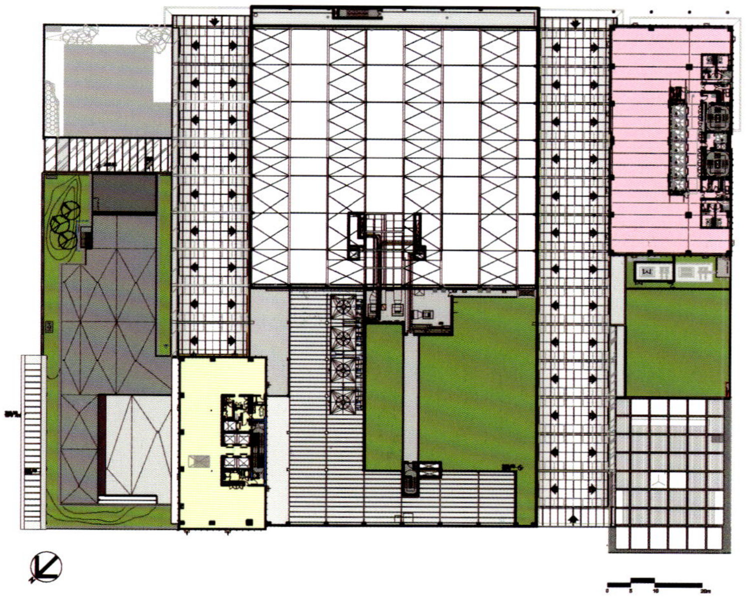

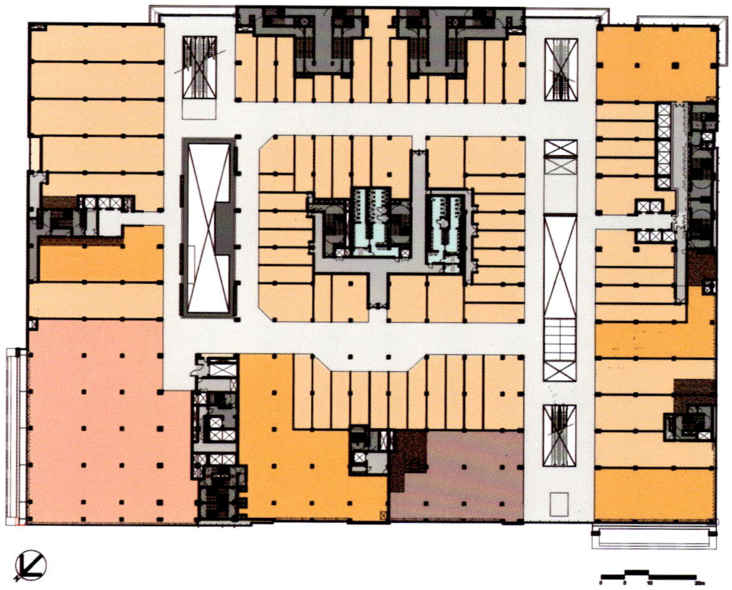

这个面积达 411975 平方米的综合开发项目包括了两个部分：一部分是新建的大型购物中心和两座塔楼，另一部分是对项目所在地址原有建筑的立面和内部进行重新设计和装修以使其变成办公楼。新建楼体的面积达 105701 平方米，包括面积为 66183 平方米的 Iguatemi 购物中心、一座 22 层的办公塔楼、一座 19 层的办公楼和该项目用户专用的地下停车场。

而对面积为 88630 平方米的现有建筑的改造和装修工程包括楼体立面的重新设计、内部平面布置和装修以及大厅和公共区域的改造。桑坦德银行是这座被改造的办公楼的常驻用户。另外一座被改造的大楼（B 座）面积为 32000 平方米，将被用作办公大楼和剧场。

UNIQA Vital Business Center Budapest

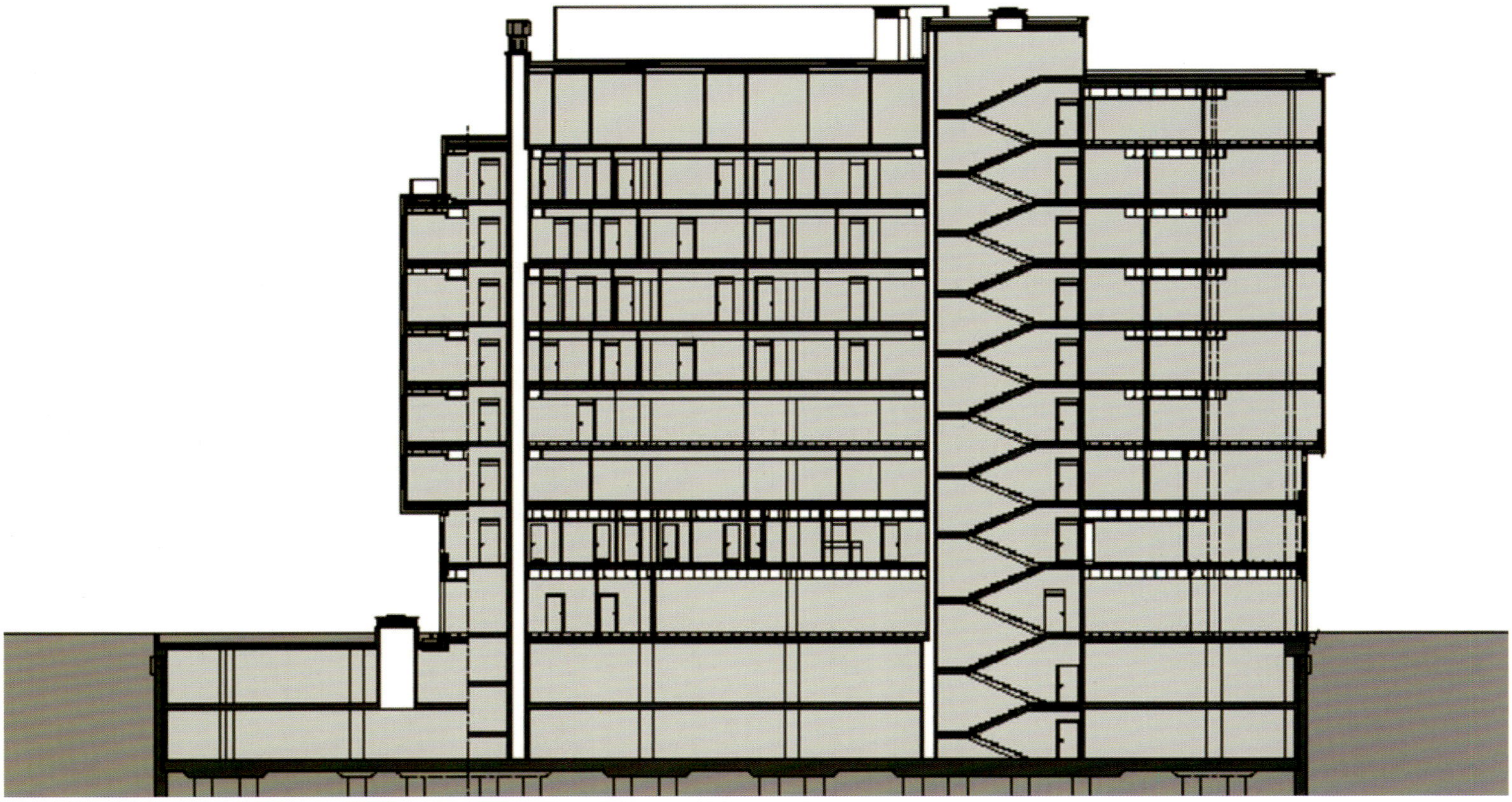

LOCATION\Hungary

The new head office is located close to the previous one at the junction of two important traffic arteries in District XIII, Lehel út and Róbert Károly körút. The office building constructed by property developer Raiffeisen Evolution in 22 months, named the Vital Business Centre, has net usable floor space of 18,000 square metres.

The nine-storey building houses all of UNIQA's divisions, while 40% of the usable floor space will be let to other companies. Companies renting the space so far include a pharmacy and a diagnostics centre.

While designing the building the architect was guided by the principle of transparency. Transparency is one of the cornerstones of UNIQA's business philosophy, and is conveyed in the Vital Business Centre by many glass surfaces and expansive rooms.

Lighting up the night

A distinctive feature of the building is its LED facade, allowing colourful images and messages to be displayed on the street-facing side of the office building every evening. Although the idea is not new, it is unique in its scale: with some 80,000 pixels the Vital Business Centre currently boasts Europe's largest LED facade.

ARCHITECT
Ferdinand and Ferdinand Architects

UNIQA
Posta
UNIQA

UNIQA
UNIQA

UNIQA

新总部位于旧总部附近，处于第十三区两个重要交通要道 Lehel 公路和 Róbert Károly 大道的交界处。办公大楼由房地产开发商 Raiffeisen Evolution 在 22 个月内完成，命名为“重要商业中心”，其使用面积达 18000 平方米。

九层的办公大楼里入驻了优尼卡保险集团的所有部门，多出的 40% 的楼面空间将出租给其他公司。到目前为止，承租商包括一家药店和一家诊所。

在设计这栋大楼时，建筑师遵循了透明化的原则。透明化是 UNIQA 经营理念的基石之一，这种理念通过“重要商务中心”的许多玻璃曲面及宽敞的房间进行了诠释。

夜间照明

大楼的一个显著特点是其 LED 立面设计。当夜幕降临时，这种设计使色彩鲜艳的形象和信息都显示在办公大楼临街的一面。虽然这种设计并不新颖，但因其规模巨大而与众不同。拥有约 8 万像素的“重要商业中心”目前可称为欧洲最大的 LED 立面。

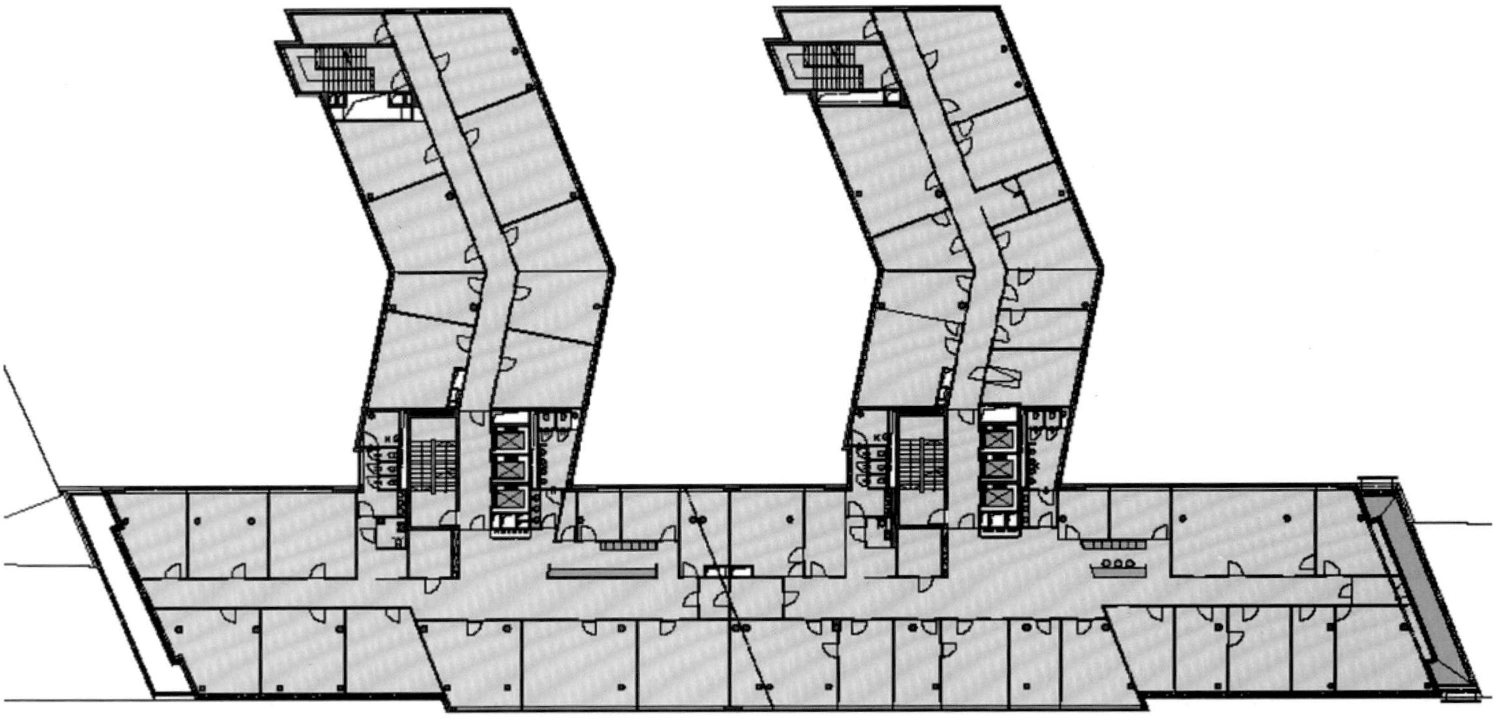

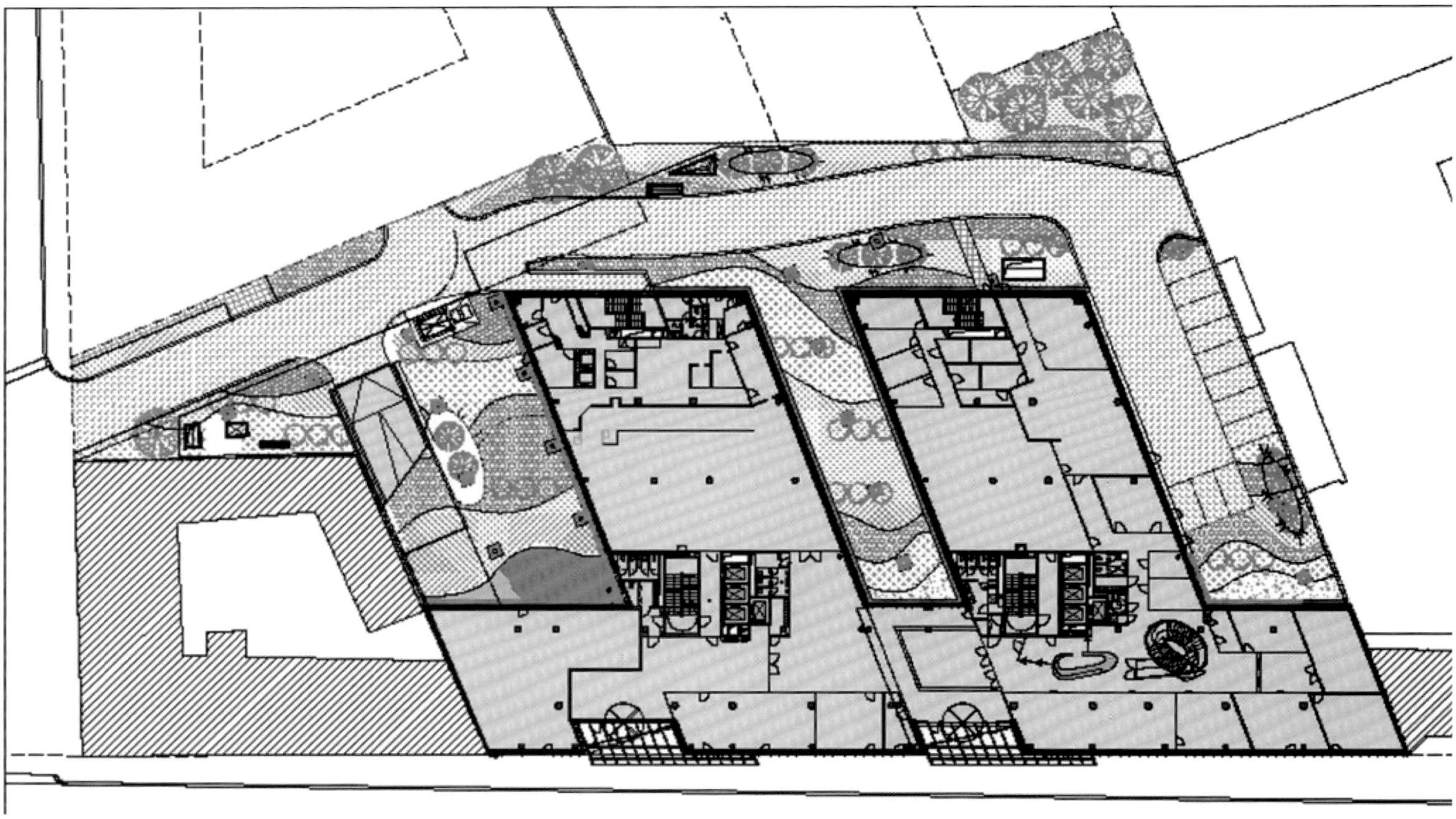

Pénztár
UNIQA

Pénztár

1100 First Street

LOCATION\1100 First Street, Washington, D.C., USA

Krueck & Sexton's architectural design vision for 1100 First Street re-defines the expectations of the speculative office building in Washington, D.C. Organized as two distinct 350,000 square-foot blocks on a 1.7 acre site, the forms of this glass-clad building pair are subtly manipulated to emphasize verticality and activate a dynamic courtyard at the centre of the complex. This space brings natural light deep into the site and identifies the building's main entry.

In a city known more for stone facades and traditional windows, 1100 First Street is decisively modern in its expression and 21st century in its technology. Glass, the building's exterior material, is used in two different but related ways: cleanly detailed and folded at the courtyard in a manner that lightens each volume and clearly identifies the main facade, and more deferential and modular at the adjacent streets. The architectural language that results from the precise articulation of surfaces and edges is timeless and enduring.

A high-performance facade, which uses glass in varying directions for careful infiltration and controlled reflection, provides for an open & light-filled building offering daylight and views to over 75% of its occupied areas. Throughout the building's office spaces, natural light penetrates deep into the plate from the floor to ceiling exterior glazing. All of the glazing units are insulated with a low-e coating, providing superior energy performance.

At the ground floor the lobby is revealed, opening up the building to the scale of the street. Detailed with care and sophistication, the lobby design draws upon the language of the exterior to create a distinctive identity. The first completed building of the pair is certified LEED Gold, achieved through site strategies, water savings and energy efficiency.

ARCHITECT
Krueck & Sexton Architects

ARCHITECT OF RECORD
Gensler

MEP ENGINEER
Flack + Kurtz

STRUCTURAL ENGINEER
Tadjer Cohen Edelson Associates, Inc.

LIGHTING DESIGNER
George Sexton Associates

LANDSCAPE DESIGNER
Daniel Weinbach & Partners

AREA
32981 m^2

PHOTOGRAPHER
Prakash Patel

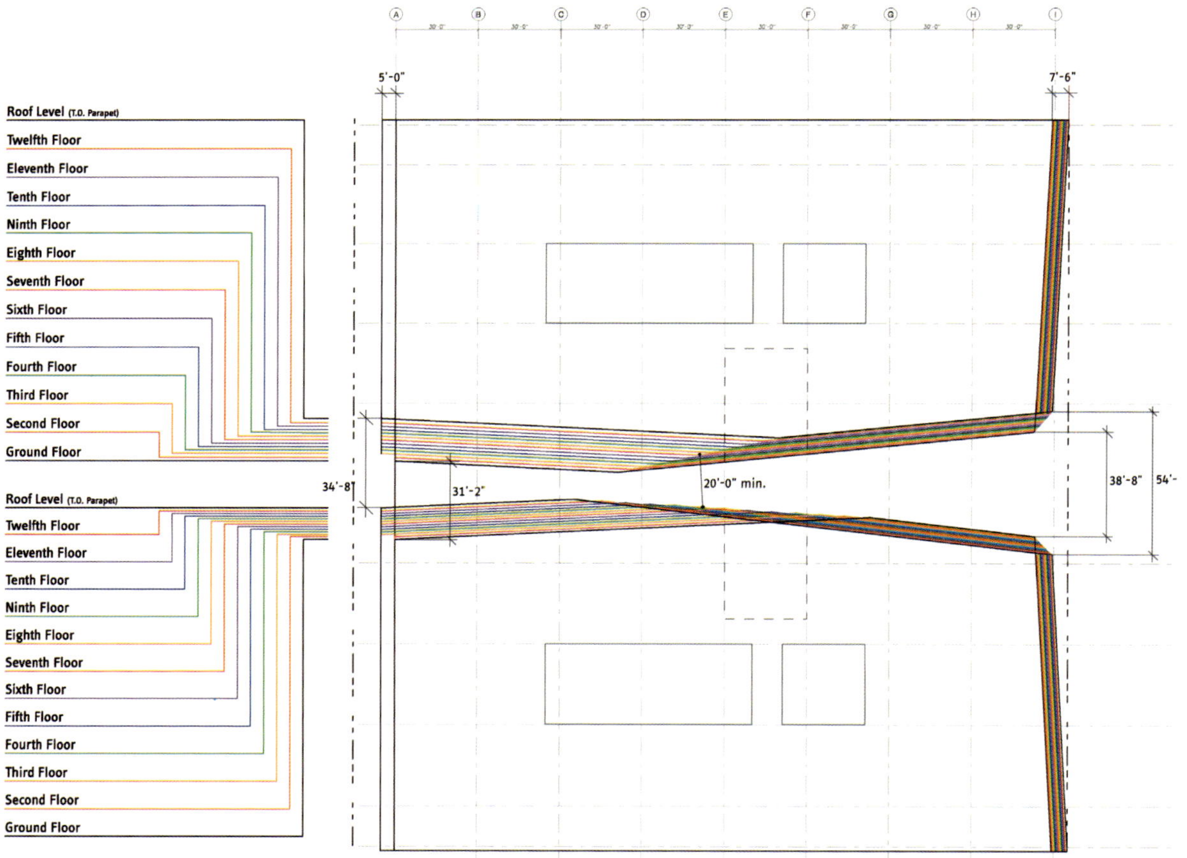

Plan Profile Diagram

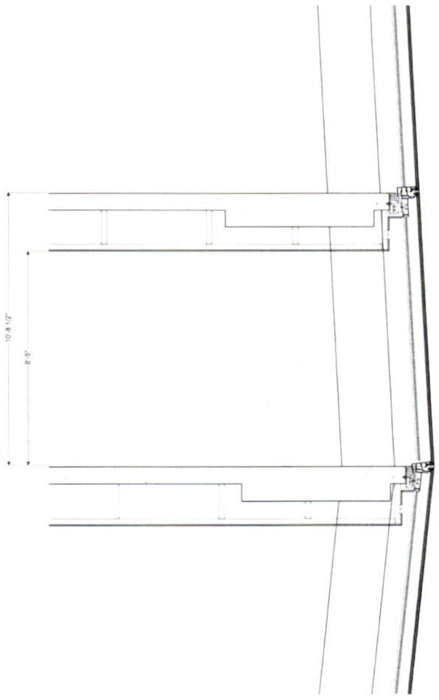

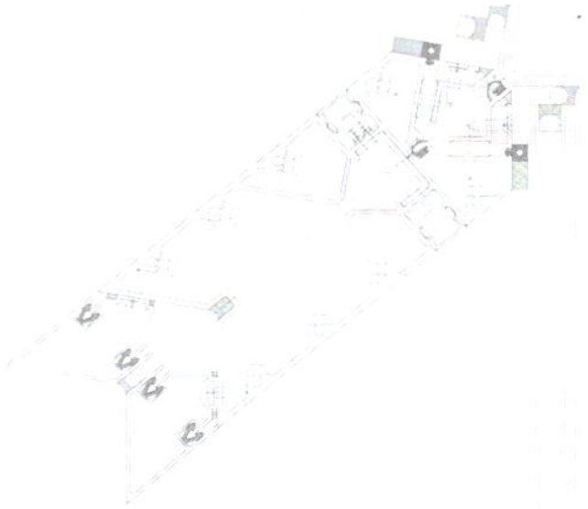

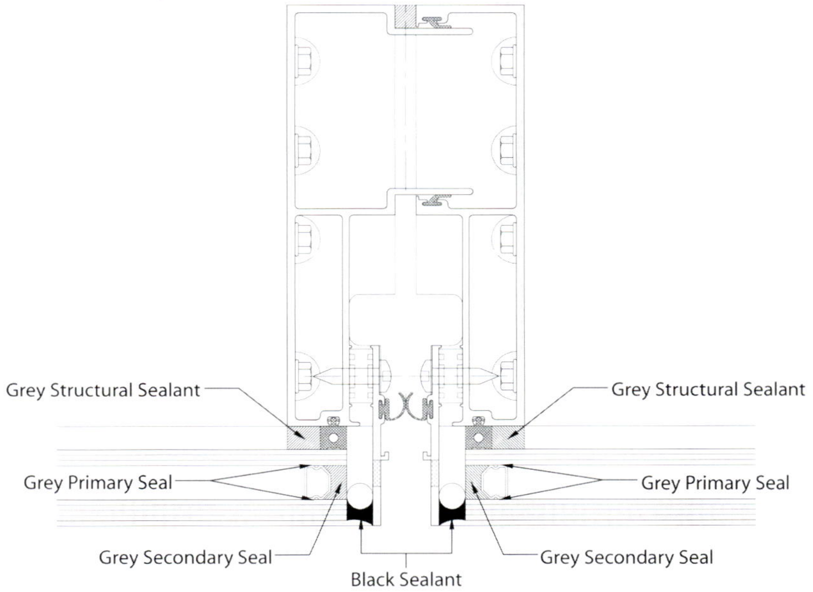

Krueck & Sexton Architects 设计的首都华盛顿第一大街的 1100 号大楼，其杰出的设计理念改变了人们对于办公大楼的固有观念。该大楼由两座建筑面积均约 32516 平方米的独特的高楼合并而成，占地面积达 6880 平方米。两座大楼的外立面都是玻璃幕墙，经过设计师的巧妙设计，特别突出了竖向垂直感，同时在两座大楼连接的正中心区域打造了一个富有生机的院落。庭院的设置使自然光线能到达大楼的内部空间深处，庭院也是大楼的主入口。

华盛顿以其石材立面和旧式窗户的传统建筑而著称，置身其中的第一大街 1100 号大楼却刻意表达现代气息并运用新世纪的新科技。大楼的外立面采用了相异却彼此呼应的玻璃材质来表现，两种形式的区别清晰而细致，并在庭院处交叉合拢。这种设计使楼内的每一个空间都充满亮光，也突出了主立面，使人从远处就能很快注意到大楼；同时也与临近街区的环境交相辉映。建筑物表面和边界的精密的交汇表达出持久和永恒的建筑语汇。

大楼的外立面利用不同方向的玻璃来调节透光率和反射率，整座大楼相对开放，光线充足，使得整座大楼的 70% 以上依靠自然光来照明。自然光透过落地玻璃窗照进整个办公区域。所有的玻璃上都有低辐射玻璃膜，绝缘性能良好，因而整座大楼有优良的能源效率。

开敞的底层门厅，使大楼对外界呈开放姿态。大厅的设计精巧而细致，每一个细节都考虑得很周全，并与楼外景观自然承接，形成自己独特的风格。第一座建成的大楼因其独特的地理环境、节水的特性和优良的能源效率，建成后即获得 LEED 绿色建筑认证。

331 Foothill Road Office Building

On the ground level, the building is accessible on all four sides and is augmented with landscape areas.

The main building entrance runs through the building on an east-west axis and continues across the entire width of the site. Along the west side of the building, a 20-foot-wide pedestrian pathway has been provided parallel to Foothill Road, not only providing building access but also an intimate green space complete with outdoor bench seating and landscaping.

Along the north, south and east façades of the facility are exterior patio areas enclosed by landscape planters. The second, third and fourth floor levels contain commercial lease space and are nearly identical in footprint, with the exception of the fourth floor level, which will also house the City's Cable TV facility.

The floor plates are enclosed by either a glazed curtain-wall assembly or a single-skin façade system with an external louver sunscreen device along the south, east and west elevations, where sun shading is necessary. The modularized screen system is made of horizontal aluminum louvers in varying widths, suspended from the floor and roof deck edges on the exterior.

The Cable TV facility on the fourth floor requires that there be no natural light or windows to the exterior, thus the building envelope surrounding this area consists of a ventilated, fiber-reinforced cement panel rain-screen assembly.

The north side of the building is primarily glass with open exterior balconies that take advantage of the indirect light and views down to the existing landscape feature.

The finishes of the interior building areas consist primarily of suspended acoustical ceiling tiles, painted gypsum wallboard and carpeted floors with the exception of all mechanical rooms, washrooms and the Cable TV facility which will have acoustical wall paneling.

The main entry lobby features a staggered strip light pattern in the bead blasted and honed stone floor, reflected in the ceiling above. Also arranged in a staggered pattern are vertical grain wood veneer panels accented with aluminum strips, covering the lobby walls.

The facility is LEED Gold certified by the U.S. Green Building Council.

ARCHITECT
Ehrlich Architects
AREA
6782 m²
PHOTOGRAPHER
RMA Architectural Photographers

LOCATION\Beverly Hills, CA, USA

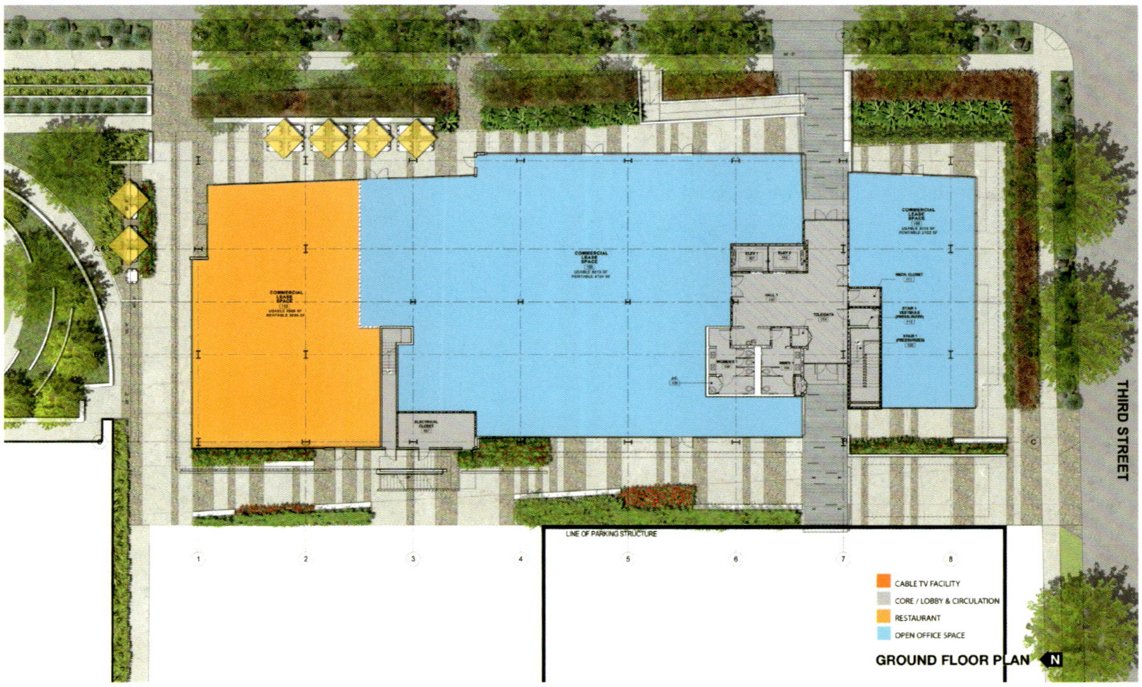

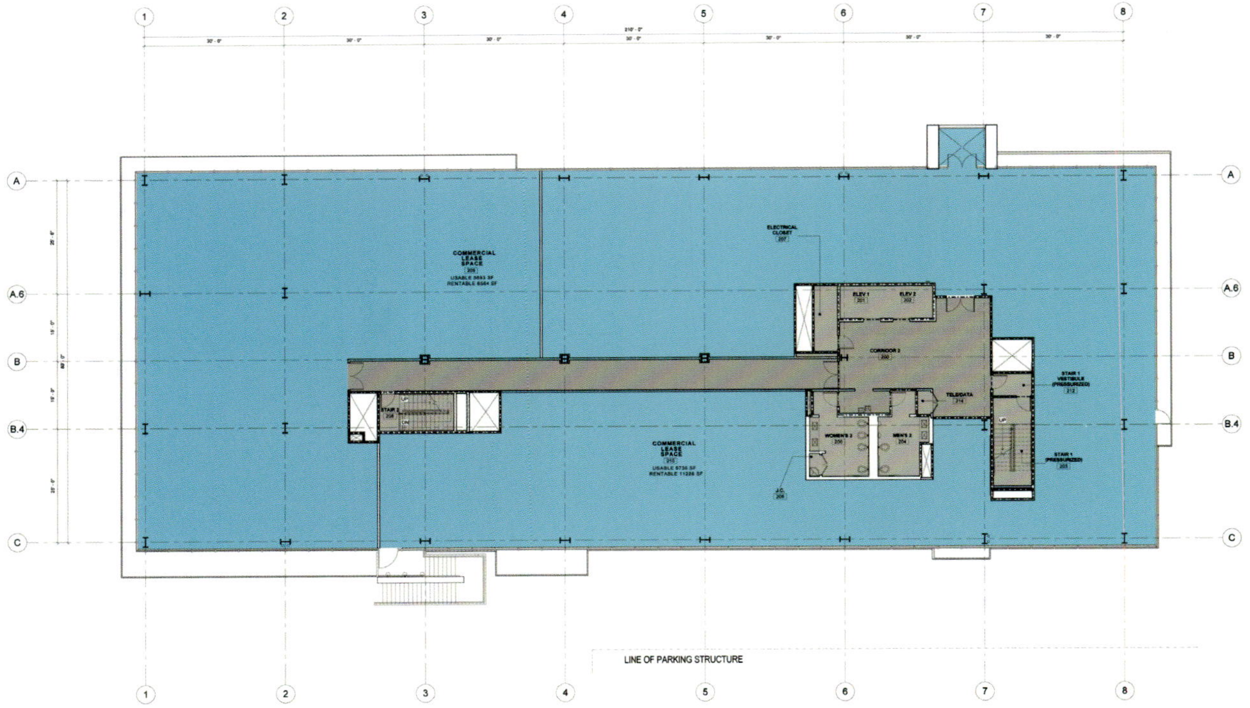

CABLE TV FACILITY
CORE / LOBBY & CIRCULATION
RESTAURANT
OPEN OFFICE SPACE

SECOND FLOOR PLAN N

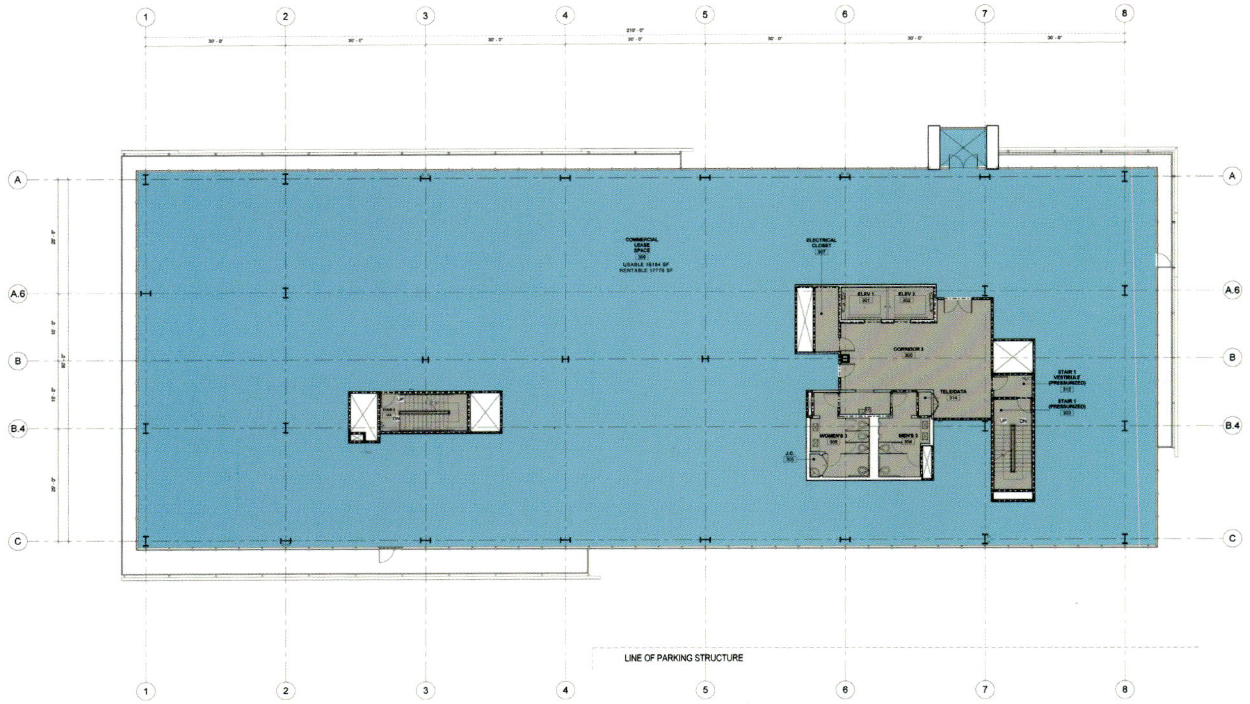

CABLE TV FACILITY
CORE / LOBBY & CIRCULATION
RESTAURANT
OPEN OFFICE SPACE

THIRD FLOOR PLAN N

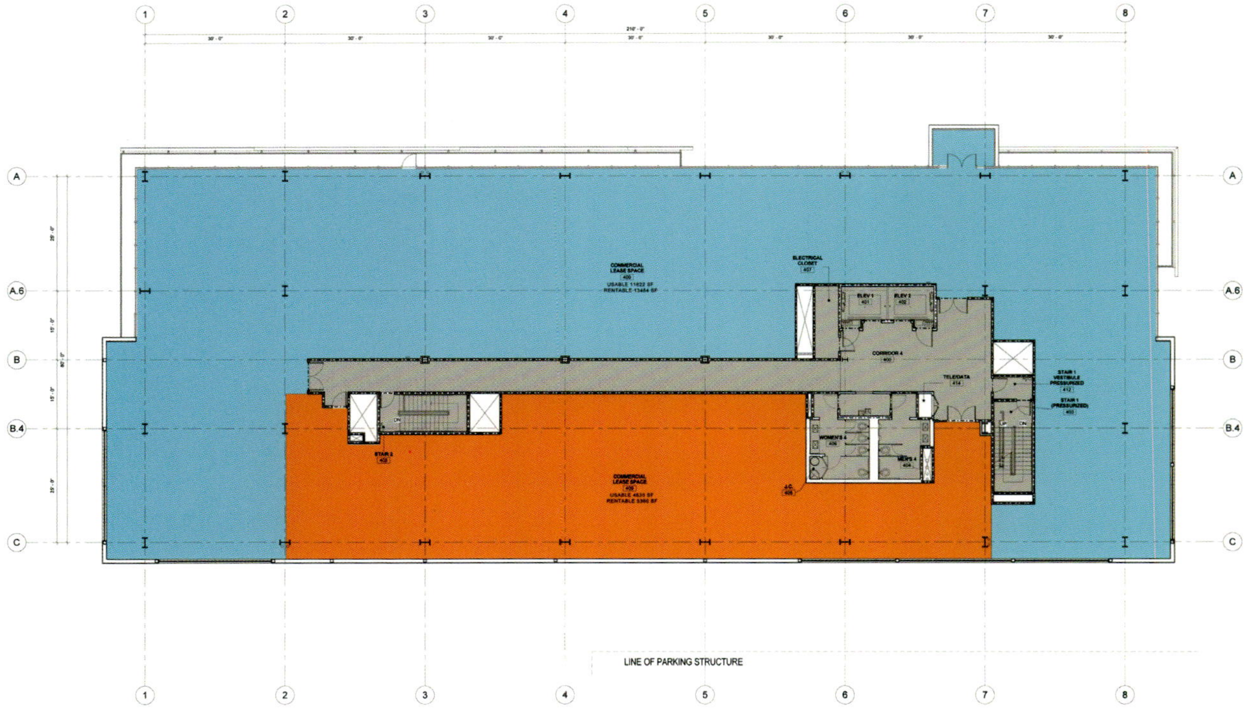

CABLE TV FACILITY
CORE / LOBBY & CIRCULATION
RESTAURANT
OPEN OFFICE SPACE

FOURTH FLOOR PLAN N

331

福特希尔路331号写字楼一楼的四面都有入口，这大大增加了项目的景观面积。

大楼的主入口沿东西轴线横穿大楼。大楼的西侧是一条与福特希尔路平行的宽6米的人行道，人行道紧挨怡人的绿化空间，其间布满景观植物和供行人小憩的长椅。大楼的西入口也在人行道上。

一楼的北、南和东三面均有户外露台，摆放着各种盆景植物。二、三和四层分布有商业出租区，楼面面积除了四层外也几乎是相等的。四层还将安置城市有线电视设备系统。

大楼部分楼板外覆盖的是釉面幕墙组件，其他部位安装的是单层玻璃幕墙系统。在大楼的北、东和西三个立面还必须考虑遮阳功能，因而在这三个立面的单层玻璃幕墙外还配有室外百叶遮阳设施。预制遮阳系统是由宽窄不一的水平铝质遮光格栅串成，从屋顶楼板和建筑外立面悬垂下来。

安放在四层的有线电视设备系统要求不能有任何自然光进入，因而四层的外墙上不能安装玻璃窗，相应区域的围护材料由透风的钢筋混凝土制成的雨幕组件组成。

大楼的北立面主要是玻璃幕墙，各层都有户外露天露台，充分利用间接光线，并可以欣赏到户外的景观。

大楼内部的装修主要采用吸声吊顶天花板以及刷漆的石膏墙板，除了厕所和机械机房外，其他地方还铺设了地毯。有线电视设备间还装上了吸声墙板。

独具特色的门厅设计引人入胜。门厅地面上铺有不锈钢喷砂和亚光石地板，地板里交错镶嵌着条形灯，地板灯光反射到头顶的天花板上，使门厅显得非常明亮。门厅四面的墙面从上到下覆盖有交错的薄木板，木板边缘用铝条加固，结合灯光创造出非同寻常的视觉效果。

该大楼获得了由美国绿色建筑协会颁发的LEED绿色建筑认证。

TELEPHONE/DATA
ROOM
EXIT
EXIT

BWT Offices

Planning specifications for the BWT Budaörs project were for a clear and concise office, sales, production, and storage building effective for advertising purposes.

The functions were accommodated in 3 specific main bodies. The tower-like office building can also be seen from a distance. The subtly differentiated façade design emphasizes the variability of the main bodies by using different glazing systems and glass colors, aluminum panels and finery surface in the area of the potential expansion.

The urban position in street axis and the architectural effect of the buildings draw the attention the customers hoped for.

ARCHITECT
Skyline Architekten
AREA
2300m²(offices)+4700 m² (storages/sales)

BWT

BWT
BWT

位于 Budaörs 的 BWT 项目在规划的时候就定位为集办公、销售、生产和仓储为一体的建筑，线条清晰简洁，建筑本身就应具有广告效应。

上述所有功能在三座主要楼体里实现。其中塔形办公大楼从很远的地方就能看得见。建筑的外立面安装了不同的玻璃系统，使用了不同颜色的玻璃和铝制板材，在可能扩建的区域也设计了精美的饰面，这些设计强调了主楼的变化。

该项目位于街道的轴线上，其独特的地理位置和别具一格的建筑效果很容易引人注意，这正满足了项目投资人所希望达到的效果。

Eastern Harbor International Tower

LOCATION\1080 East Daming Road, Shanghai, China

As the winning entry of an invited design competition, this high-rise office building accommodates a multi-use building project of 40,000 square meters located two blocks from the Huangpu River in Shanghai. With dramatic views of Pudong and Bund districts, this building boasts an office tower with a rooftop garden 120 meters high. The building design is oriented to the skyline view by orienting the southwest corner of the tower toward the river and creating a glass bay and terrace at the uppermost floors. The mixed-use tower sits atop a 21 meters high podium that houses a health club, pool, conference facilities and ground floor retail space. A restaurant and terrace are located on the top floor of the office tower allowing for commanding views of the skyline. Adjoining this mixed-use office complex, a 10 meters tall building is dedicated to retail use. Extensively landscaped outdoor areas and a grand lobby create an inviting environment that reflects the openness and flexibility which are key concepts of the design. Unimpeded circulation throughout these spaces is also a main consideration for the project.

The design incorporates a favorable site and pleasing indoor environment with essential sustainable features, such as energy and water conserving measures, recycling, and renewable materials that enabled it to pursue LEED Gold Certification. While sustainability is a key aspect of the building process, all design and construction materials fulfill fire protection and security consistent with the National Standards of China.

ARCHITECT
Heller Manus Architects, Jeffrey Heller, FAIA, Partner-in-Charge

AREA
40000 m²

SUSTAINABLE FEATURES
LEED-CS Gold Certified (First LEED Gold high-rise in Shanghai)

PHOTOGRAPHER
Heller Manus Architects

位于上海的东方海港国际大厦，总面积达 4 万平方米，与黄浦江仅隔 2 个街区。作为设计邀请赛的获奖作品，高 120 米的办公塔楼将浦东及外滩的美景尽收眼底，楼顶花园更显其奢华。塔楼的西南角朝向黄浦江，在最高层打造有玻璃开间和露台以开辟临江天际线景观。多功能塔楼的底部是一座高 21 米的裙楼，裙楼为商业楼，内设康体俱乐部、游泳池和会议设施，一层还开辟有商业空间。餐厅和露台设在塔楼的顶层，可以饱览天际线景观。多功能写字楼群附近有一幢 10 米高的建筑，主要用于零售业。大堂入口与室外景观的自然承接突显了开放与布局灵活的设计理念。设计还考虑到各个空间之间的畅通。

设计将节水、节电、回收、再利用等根本的可持续发展特色融入到宜人的室内环境中，结合项目的有利地理位置和周围的环境，使之具备角逐 LEED 绿色建筑认证的能力。该项目在建设过程中不仅高度重视可持续性发展，而且所有的设计和建筑材料均符合国家消防和安全标准。

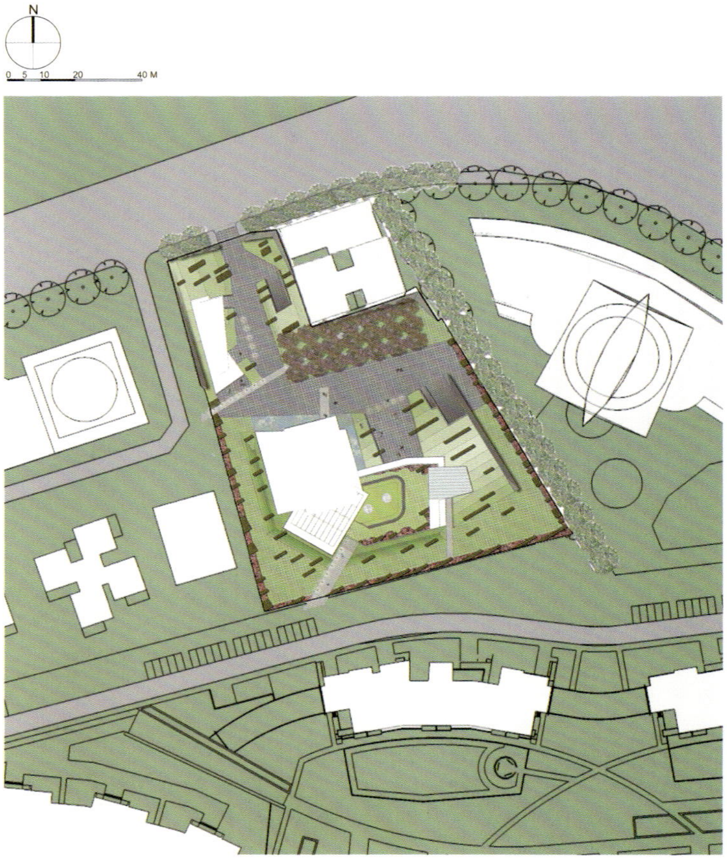

SITE PLAN

LEASING
OFFICE

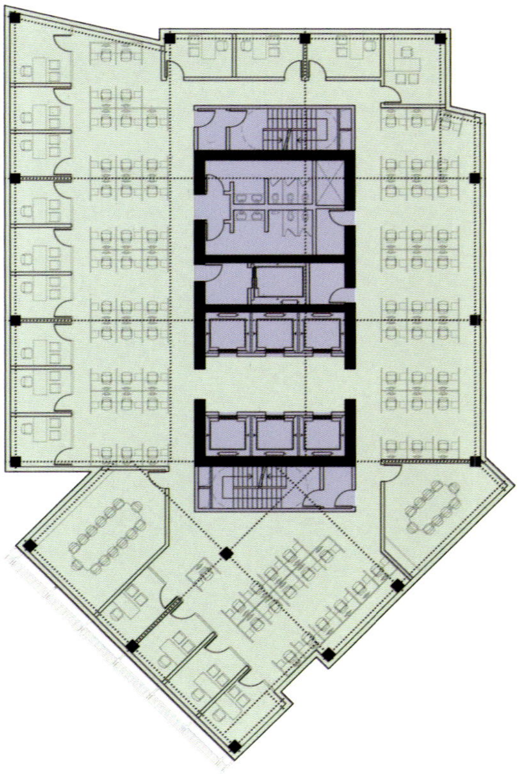

TYPICAL LEVEL FLOOR PLAN

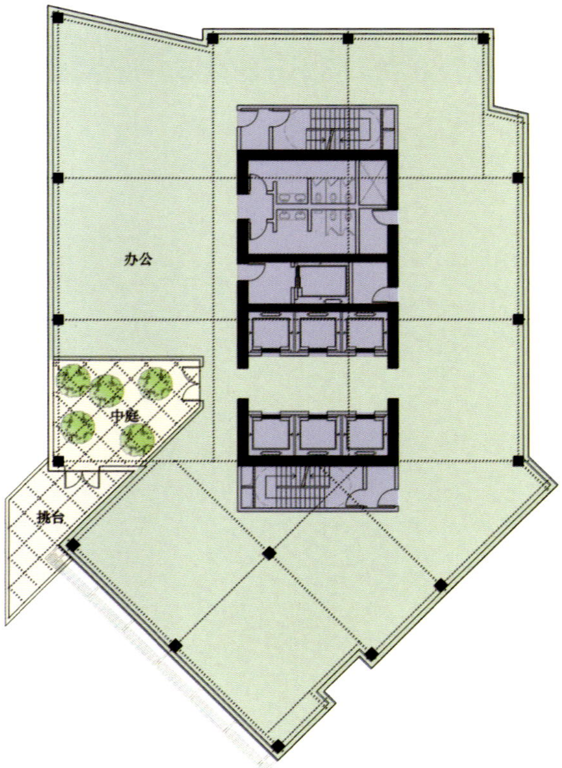

22TH-26TH LEVEL FLOOR PLAN

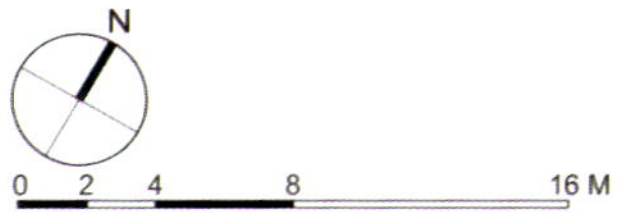

Oficina Vidre Negre

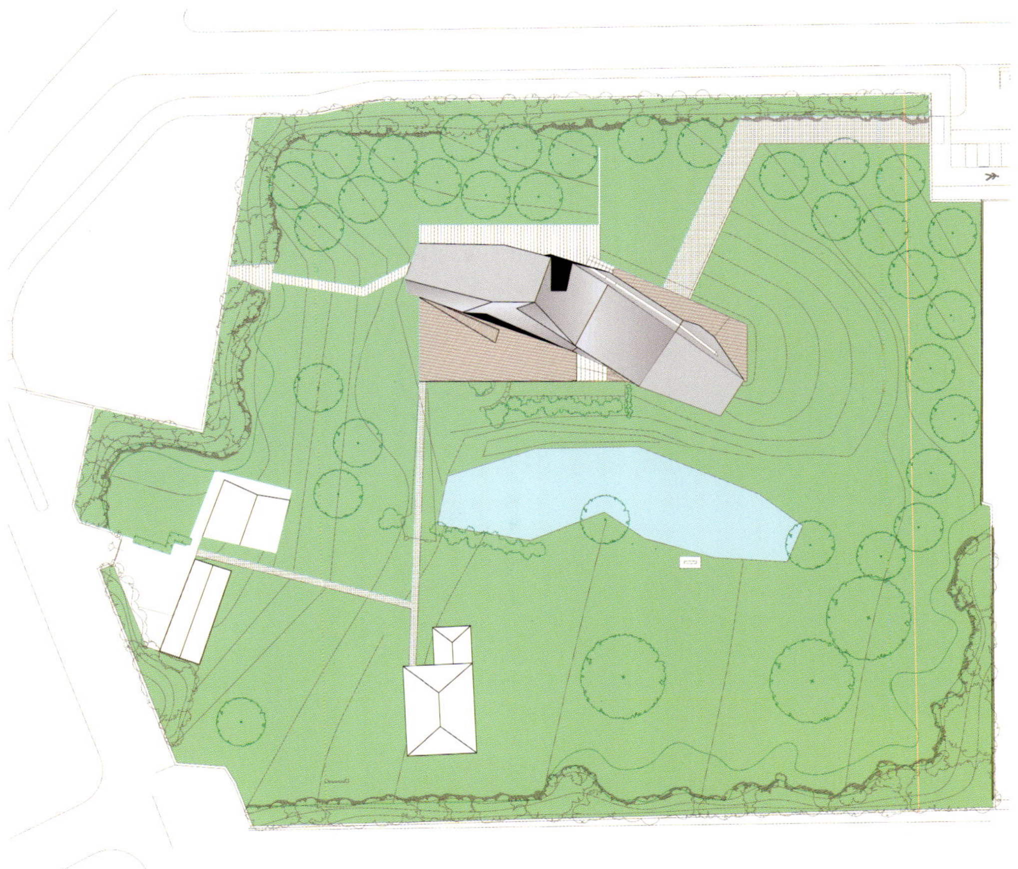

LOCATION\Piemonte, Italy

Oficina Vidre Negre was born as a contemporary sculpture, symbol of a dynamic and continuous development that needed strong identification. Even the building seems to be evolving. Multifaceted prism completely covered in black glass and integrated photovoltaic panels are projected on the vast expanse of water. Besides the stage of reflection and suspension of the volume, the basin is the water reservoir for the plant fire.

The breakdown of volumes is kept indoors, a succession of spaces intersected in four split-levels, illuminated by cutting higher education, and side windows. In plan the building is developed into two wings separated by a central block of facilities (stairs, toilets and bars). Operating areas are designed as open-space, while the offices more private are in the glass that is projected on the blade of water.

The entire structure is made of concrete frames with external cut aluminum and glass with low emissivity.

ARCHITECT
Duilio Damilano

FIRM
Damilano Studio Architects

COLLABORATOR
Dott. Arch. Enrico Casetta : parametric modelling
Arch. Alberto Pascale : virtual modelling
Arch. Jessica Pignatta
Arch. Claudia Allinio
Arch. Manuela Taricco
Arch. Rodolfo Cerati
Dott. Arch. Emanuele Meinero

AREA
6960 m^2

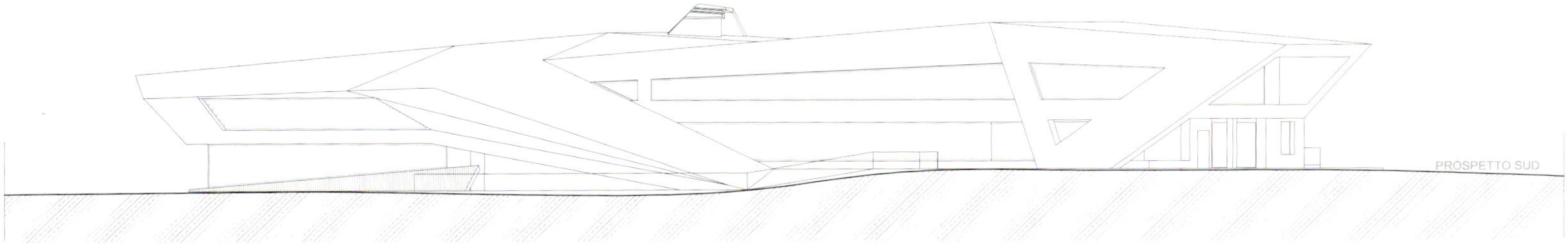
PROSPETTO SUD

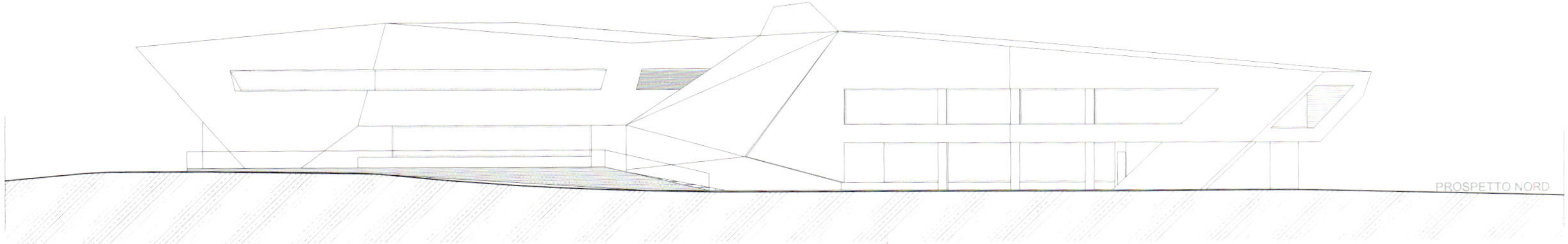
PROSPETTO NORD

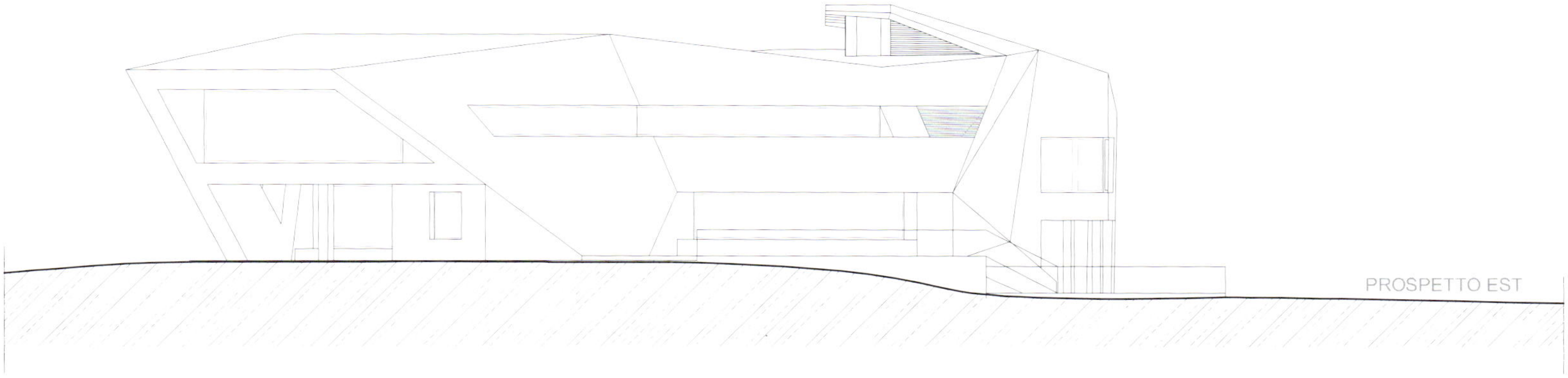
PROSPETTO EST

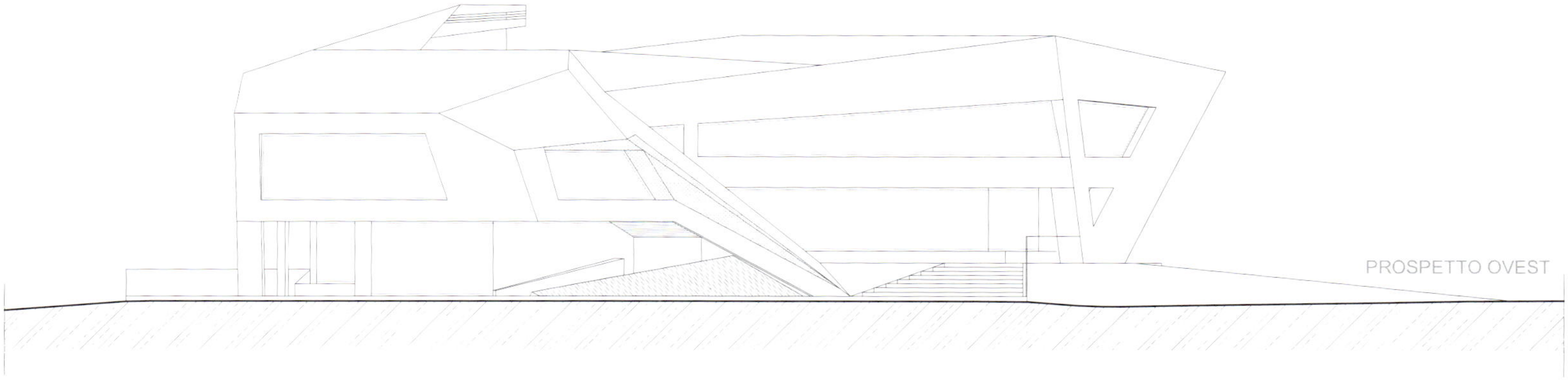
PROSPETTO OVEST

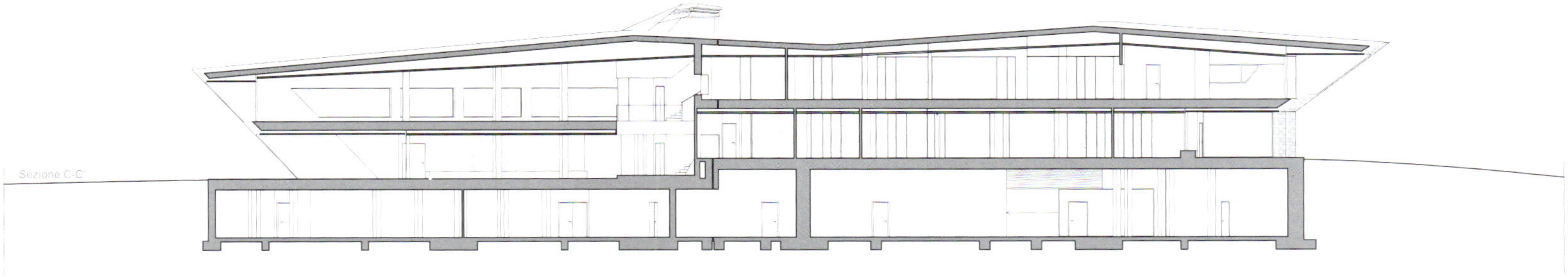
Sezione C-C

A
A

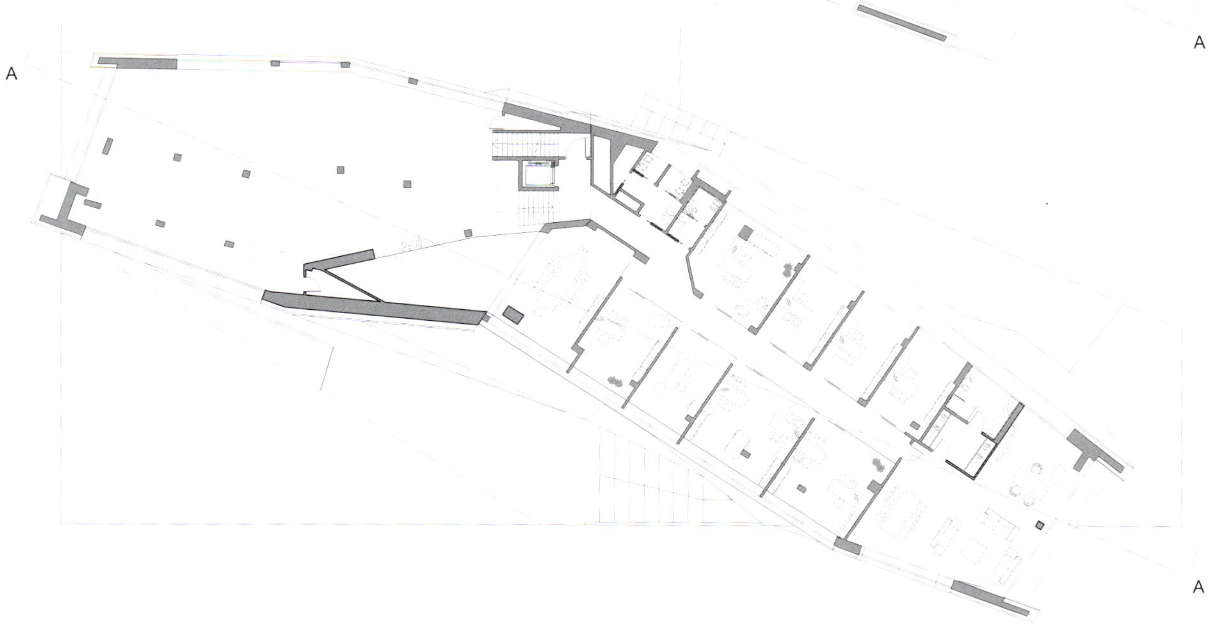
A
A

韦德里尼格拉办公楼像一座象征着活力和不断发展的现代雕塑，具有明显的可辨识性，似乎连建筑物本身都处在不断变化中。外立面犹如由黑色玻璃完全覆盖的多面棱镜，向外突出的集成光伏板投影在室外的大片水池上。在设计上，除了表现建筑物的倒影和悬浮外，水池也是火警储水的装置。

建筑物内部分解成几个部分，一系列的空间被分割成四个错层，由高科技设备和侧窗解决照明问题。整座建筑由中央设施区（内设楼梯、卫生间和酒吧）延伸出两个侧楼。工作区设计为开放空间，而较为私密的办公室则位于突出于水域上的那个侧楼中。

整个结构由混凝土框架与低辐射玻璃构成。

Breathing Factory

ARCHITECT
Takashi Yamaguchi

FIRM
Takashi Yamaguchi & Associates

ASSOCIATE
Jun Murata (Takashi Yamaguchi & Associates)

STRUCTURE
Soutaro Hayashi, Steel Frame

CONTRACTOR
Makoto Construction Co., Ltd.

AREA
2288.37 m²

PHOTOGRAPHER
Takashi Yamaguchi & Associates

LOCATION\Osaka, Japan

The requested program was about the reconstruction of the factory by giving two different options. Firstly, the completion of the existing part of the building and then the construction of the whole part of the program and design was expected.

The first floor is divided into conference room, reception office, management office and warehouse. On the second and the third floors, there are gathering functional spaces like a production spaces as the fabrication department and development department which are vertically joined as a "rift" system that break the frontier between the two floors. The fourth floor was allotted to the one room with a large meeting space which can be responded to the office for design research and the development and various seminars and meetings.

The building body is covered with a delicate membrane constituted of aluminum louvers in order to avoid showing the disgusting clustered pipes. Concerning the maintenance ease, the louver aperture ratio and the space behind those offer a proper access to the pipes system.

The reconstruction was also an experimental way to reduce as much as possible the impact on the neighborhood of the intimidating volume of the uninteresting factory. To attain to this result, the louver's angles, and horizontal or vertical directions are directed by randomized mathematical rules.

Some of the horizontally directed louvers are reflecting the moving clouds in the blue sky or the cars light crossing the street into the night. By moving the point of view, by looking at some of the vertically directed louvers, the segmented perception of the neighborhood's image is changing. By reflecting the changing nature and vicinity scene of the building, this effect is fragmenting the building and his environment perception while preserving a form of human presence feeling.

The same principle goes for the void "light court" linking the relaxation spaces on the third and fourth floors. By using the glass screen inserted into the verticals and horizontals openings, the image of the nature enters into the inside of the building. The glass screens take, transform and bring a sort of virtual image of the everyday life from the outside into the monotony filled inner space of the factory.

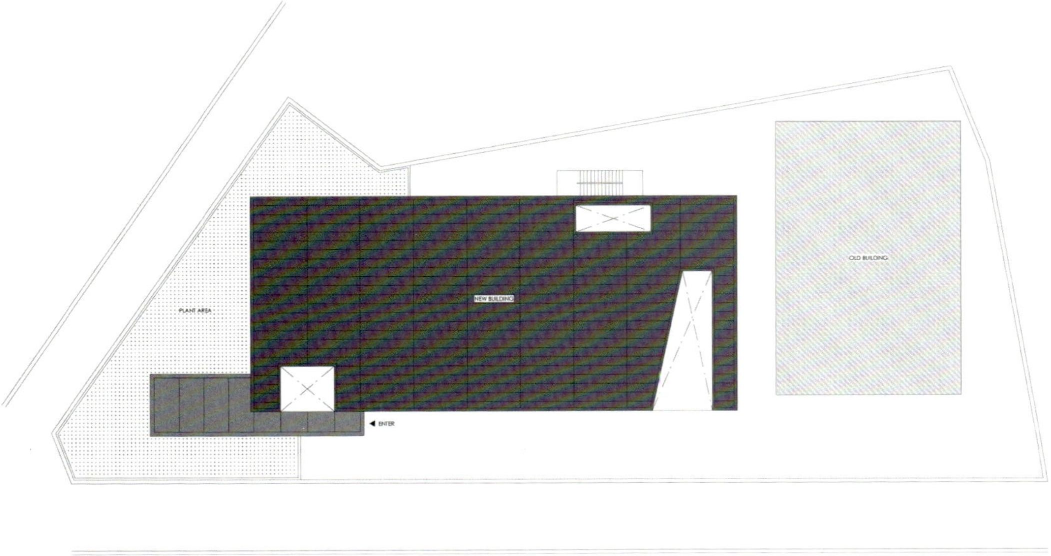

SITE PLAN
S=1/400

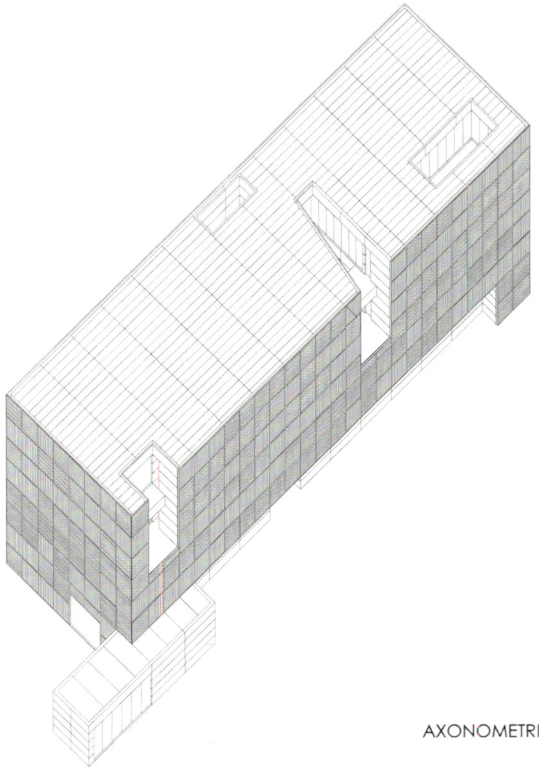

AXONOMETRIC

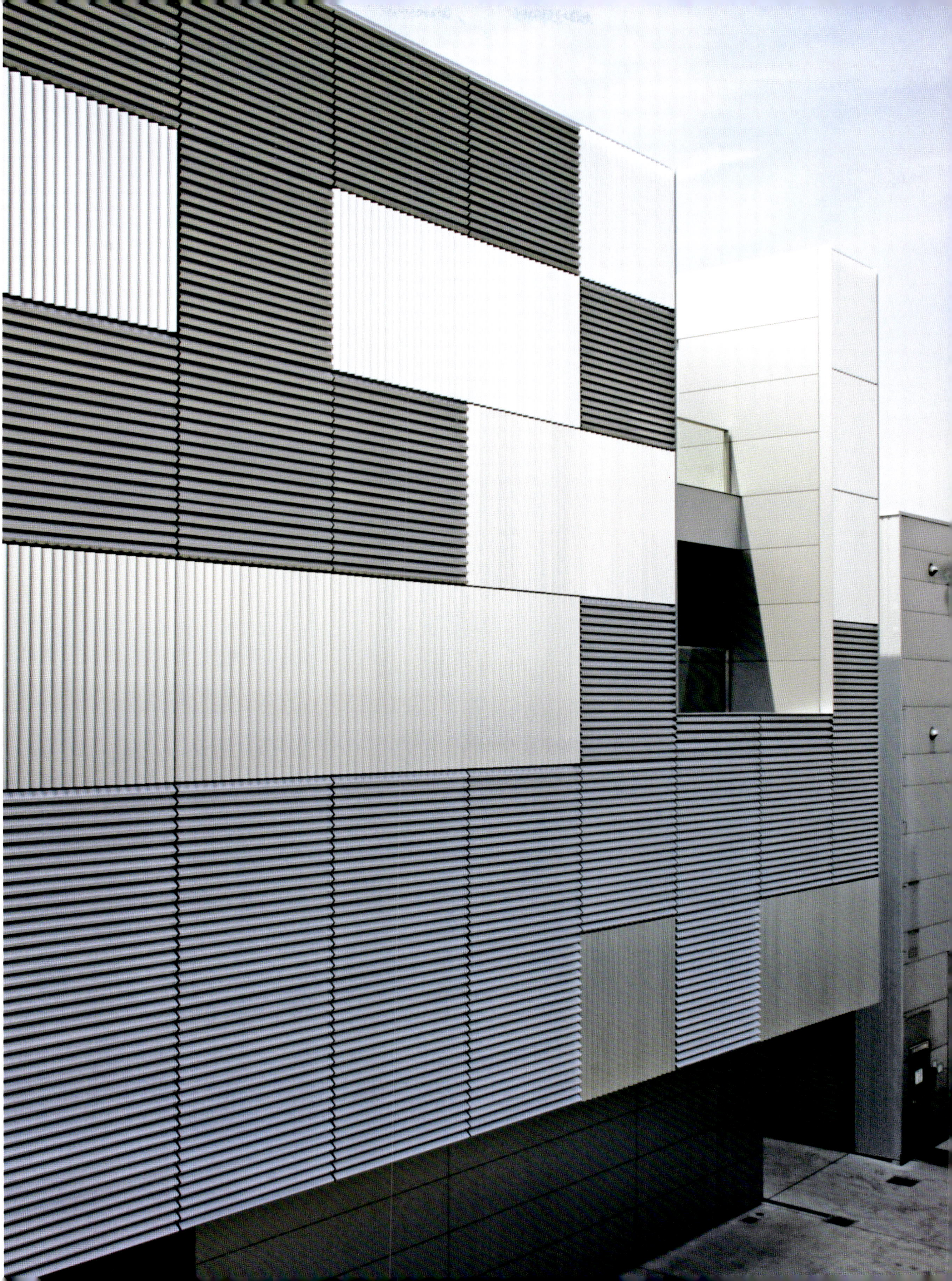

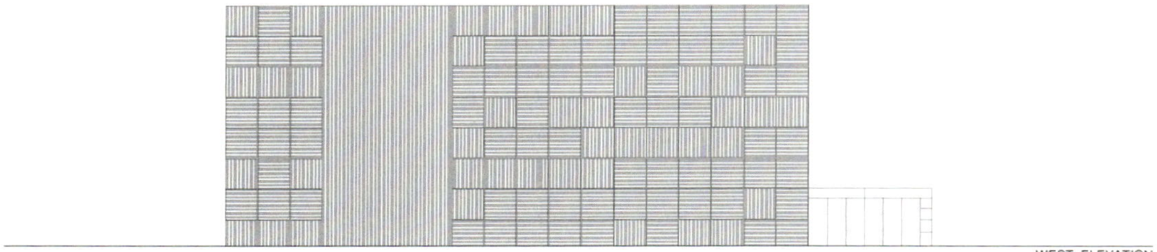

WEST ELEVATION

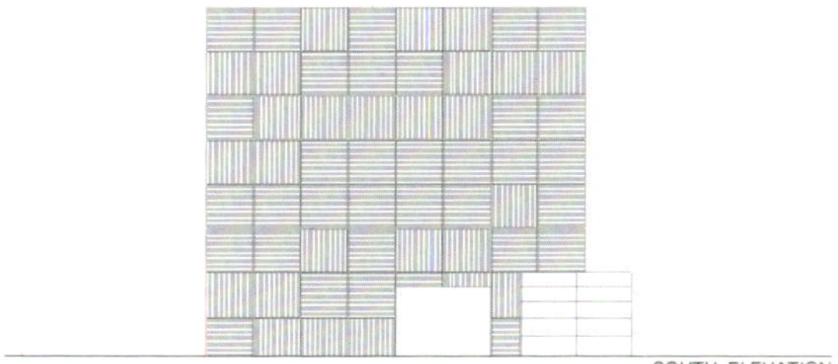

SOUTH ELEVATION

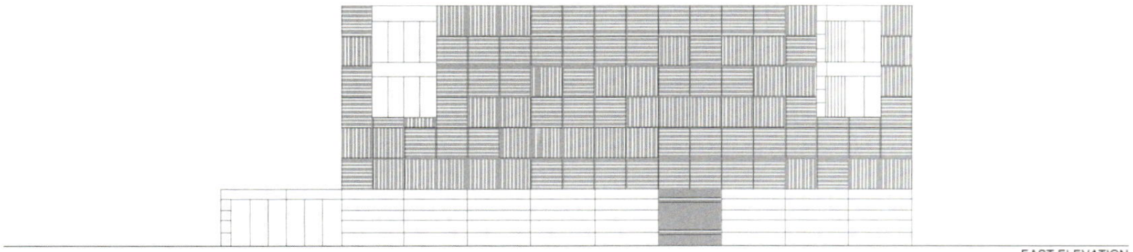

EAST ELEVATION

NORTH ELEVATION

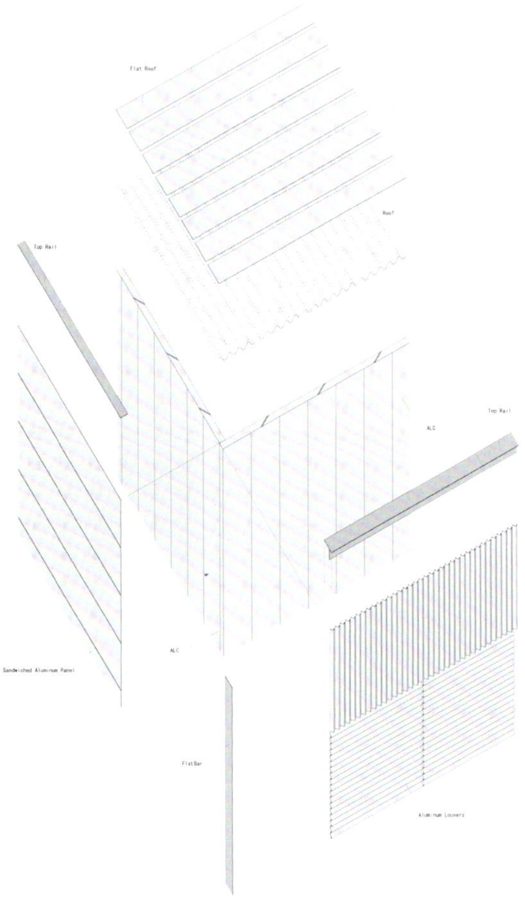

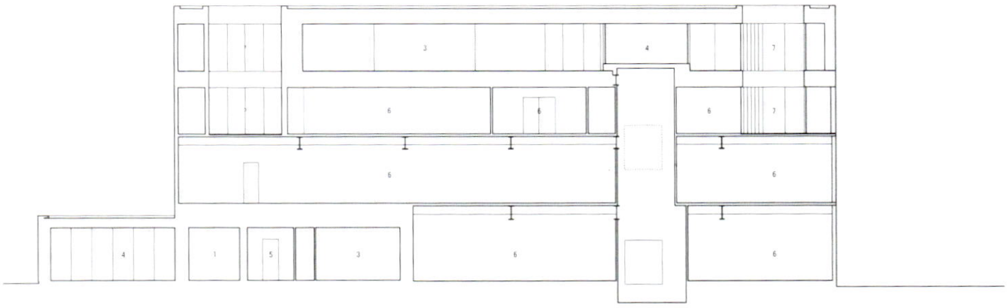

a-a' section

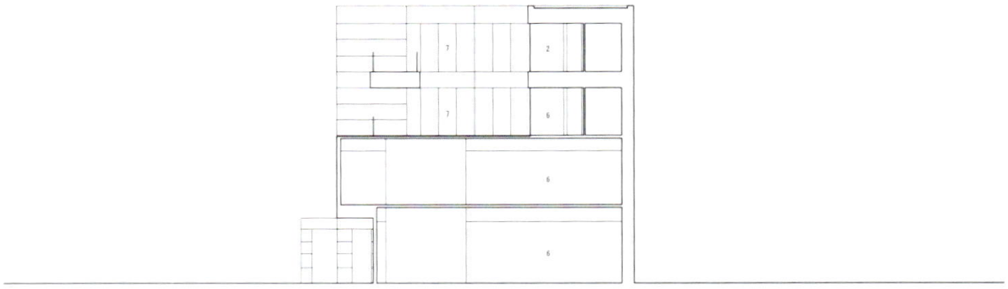

b-b' section

1. Entrance Hall
2. Foyer
3. Office
4. Conference Room
5. Utility
6. Workspace
7. Terrace
8. Storage
9. Multu-purpose Hall

Section S=1/400

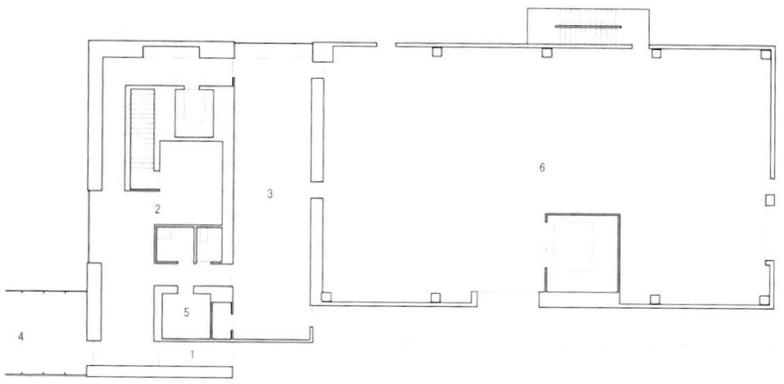

1F PLAN
S=1/300

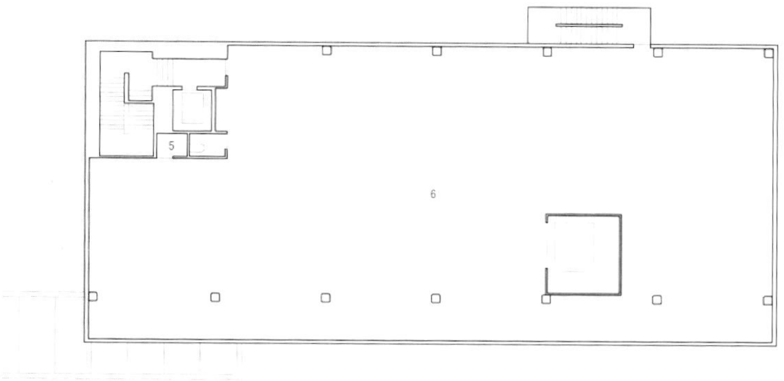

2F PLAN
S=1/300

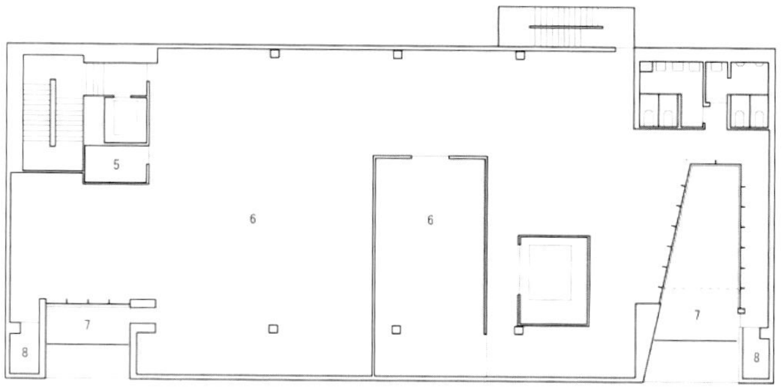

3F PLAN
S=1/300

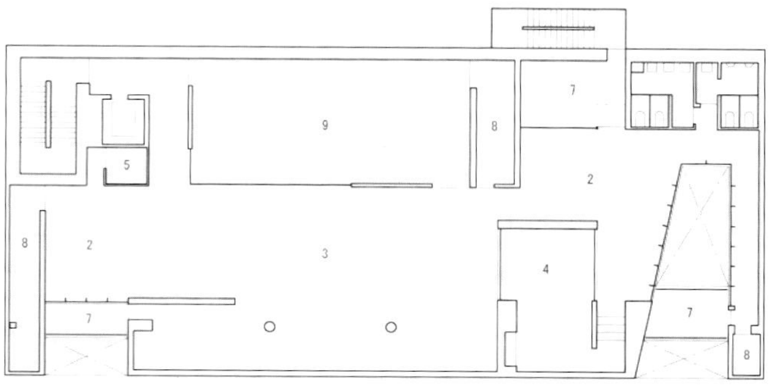

4F PLAN
S=1/300

1. Entrance Hall
2. Foyer
3. Office
4. Conference Room
5. Utility
6. Workspace
7. Terrace
8. Storage
9. Multu-purpose Hall

工厂重建项目的建设方案要求分两步实施：第一步完成对既有楼体的重建，然后再建项目的其余部分。

建筑的一层设有会议室、接待室、行政办公室和库房。二、三层为多功能空间，在这里，生产空间与制造及研发部门融为一体。整个多功能区呈纵向体系，使二层和三层没有楼层的分隔。第四层作为大型会议区，不仅可以为设计、研发部门办公使用，而且还可以承办各类研讨会议。

楼体外立面覆盖有精致的铝制百叶结构，避免了大量管线的裸露。而且百叶结构的开口比例和百叶结构的后部都预留了足够的空间，以方便维修。

此次重建也是尝试减小这个不太有趣的工厂建筑对周边环境的视觉冲击。因此，外立面百叶的角度以及百叶的摆放方式完全采用随机的数学规则。

日间，水平的百叶结构与蓝天白云相映成趣；夜间，它们会反射过往车辆的灯光。从不同的角度看去，透过纵向百叶结构，周边的视野也随之变换。该建筑不仅同周围自然景观及邻近建筑景观相映成趣，而且也分隔了楼内环境及其景观视野，从而使存在感得以保留。

连接三、四层休闲空间的天井同样也遵循了这一设计原则。水平和竖直方向的开放地带都使用了玻璃幕墙，将室外的景观引入室内。同时也将室外的日常生活景观带入单调的工厂空间。

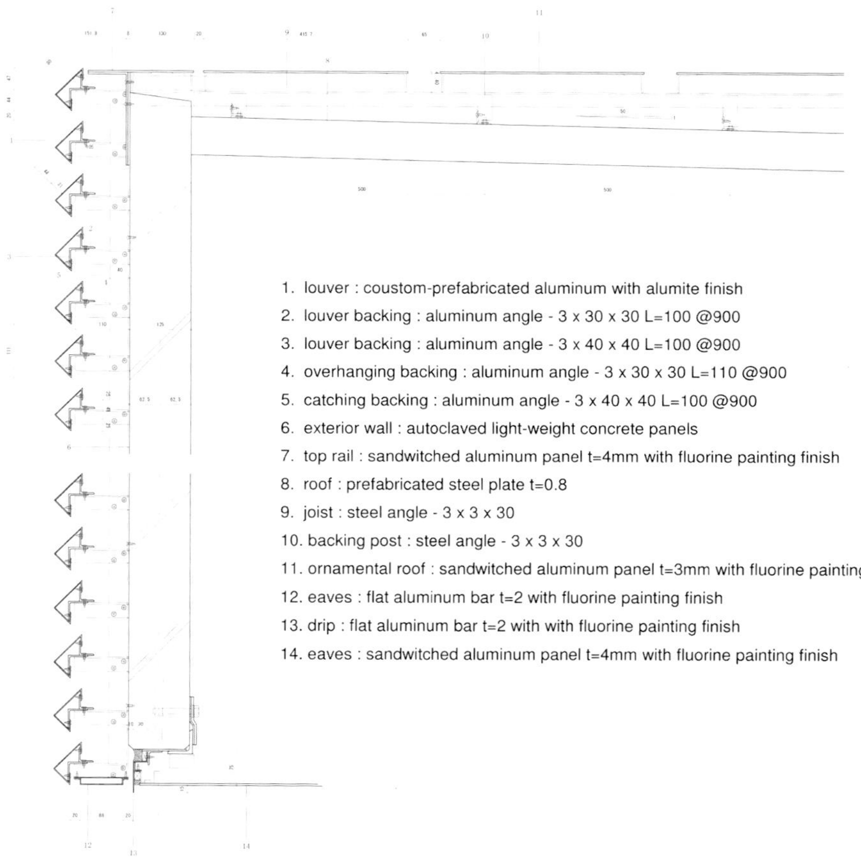

1. louver : coustom-prefabricated aluminum with alumite finish
2. louver backing : aluminum angle - 3 x 30 x 30 L=100 @900
3. louver backing : aluminum angle - 3 x 40 x 40 L=100 @900
4. overhanging backing : aluminum angle - 3 x 30 x 30 L=110 @900
5. catching backing : aluminum angle - 3 x 40 x 40 L=100 @900
6. exterior wall : autoclaved light-weight concrete panels
7. top rail : sandwitched aluminum panel t=4mm with fluorine painting finish
8. roof : prefabricated steel plate t=0.8
9. joist : steel angle - 3 x 3 x 30
10. backing post : steel angle - 3 x 3 x 30
11. ornamental roof : sandwitched aluminum panel t=3mm with fluorine painting finish
12. eaves : flat aluminum bar t=2 with fluorine painting finish
13. drip : flat aluminum bar t=2 with with fluorine painting finish
14. eaves : sandwitched aluminum panel t=4mm with fluorine painting finish

Sectional Detail of Louver and Flat roof
S=1/10

SOLON SE

ARCHITECT
Schulte-Frohlinde Architekten
SERVICE
EGS-plan Ingenieurgesellschaft, Gebäude & Solartechnik
LANDSCAPE
HochC Landschaftsarchitektur
MONITORING
IGS, TU Braunschweig
AREA
8000 m^2 (administration) + 21000 m^2 (production, research and office)
PHOTOGRAPHER
Myrzik & Jarisch, Marcus Bredt, Silke Reents

The roof of the administration area with its accessible landscaped terraces is pierced by five courtyards that bring daylight into the deep plan. The interior is arranged as a large flowing space with fire doors, which close only in the event of fire, hidden behind the stairwell enclosures. The three upper floors are terraced with open-plan office areas divided by ceiling-high cores with kitchens, lockers, changing rooms, charging stations and seating. In addition to the 360 workspaces in the administration building, there are numerous places for formal and informal discussions. Small cells allow for concentrated work and private phone calls or discussions.

The facade comprises a prefabricated timber frame incorporating triple-glazing with solar control coating, external shading fins, opening windows, and horizontal wooden slats for weather protection.

Workspaces are fully wired and include a newly developed 9W LED lamp with 250 lux performance, chargers, a retractable laptop compartment and LCD screen. Specially developed energy shuttles, charged overnight from stored roof-generated solar power or from off-peak electricity, can supply a fully-equipped workstation for ten hours. These can operate anywhere in the building, including the roof, garden or courts.

The 340mm thick concrete floors are exposed and incorporate water pipes which recycle waste heat from production areas and can be topped up in the winter by the district heating system. Solon plans to build a bio-gas plant on site which will produce all the required energy, making the building carbon dioxide neutral in operation. Five local ventilation systems are installed, corresponding to fire compartments, which also help minimize the ventilation cross sections. Mechanical ventilation is expected to be used only in winter.

Rainwater is stored in 75,000 litres underground tanks and used for toilet flushing and garden irrigation; excess rainwater is fed to filtering trenches in the landscaped garden. A solar filling station, feeding ten electric scooters for use by employees and visitors, is supplied by a double-axis solar panel with battery storage. On a sunny day up to 1,000km mileage for the motorcycles is supplied.

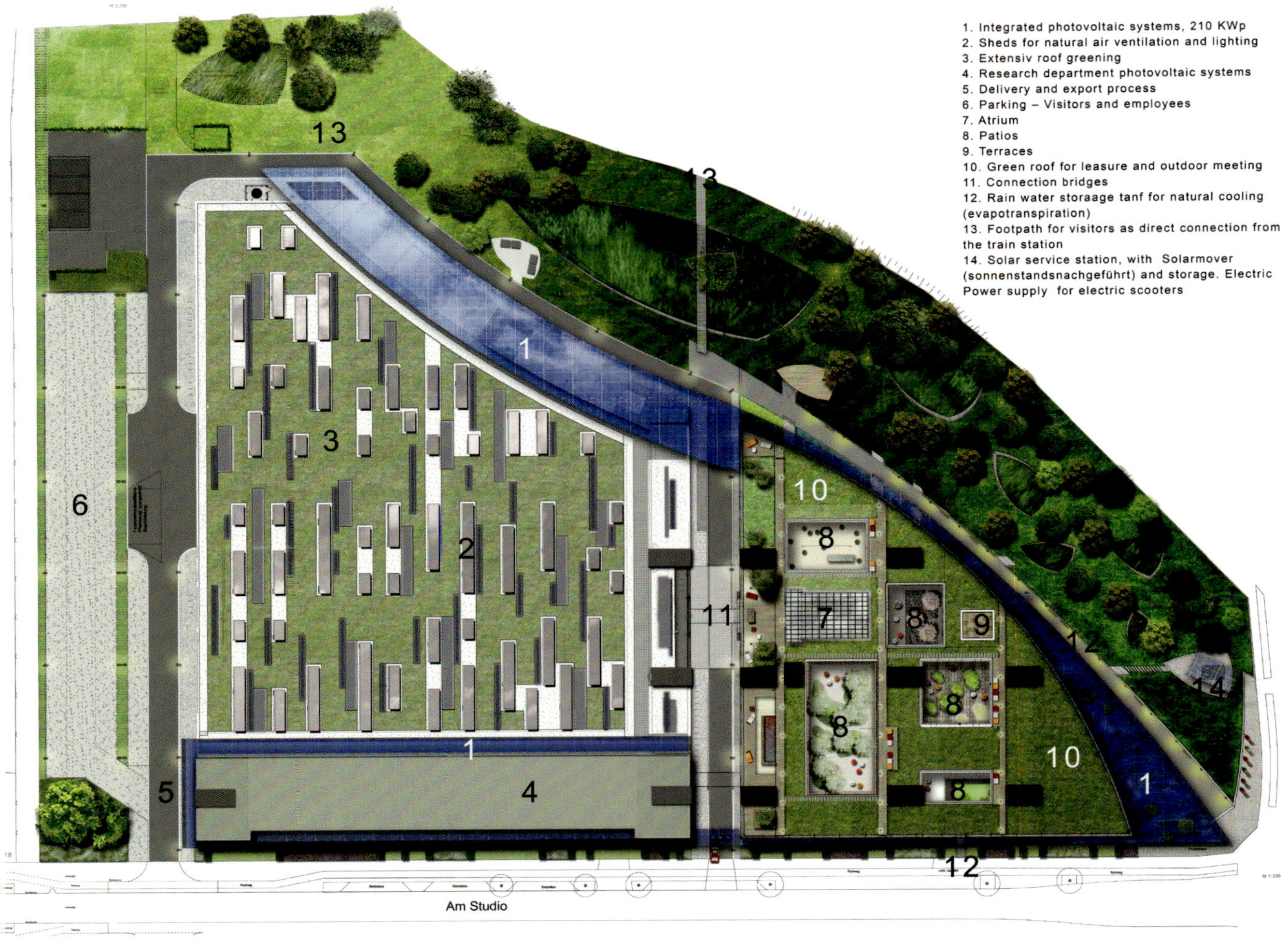
1. Integrated photovoltaic systems, 210 KWp
2. Sheds for natural air ventilation and lighting
3. Extensiv roof greening
4. Research department photovoltaic systems
5. Delivery and export process
6. Parking – Visitors and employees
7. Atrium
8. Patios
9. Terraces
10. Green roof for leasure and outdoor meeting
11. Connection bridges
12. Rain water storaage tanf for natural cooling (evapotranspiration)
13. Footpath for visitors as direct connection from the train station
14. Solar service station, with Solarmover (sonnenstandsnachgeführt) and storage. Electric Power supply for electric scooters
Am Studio

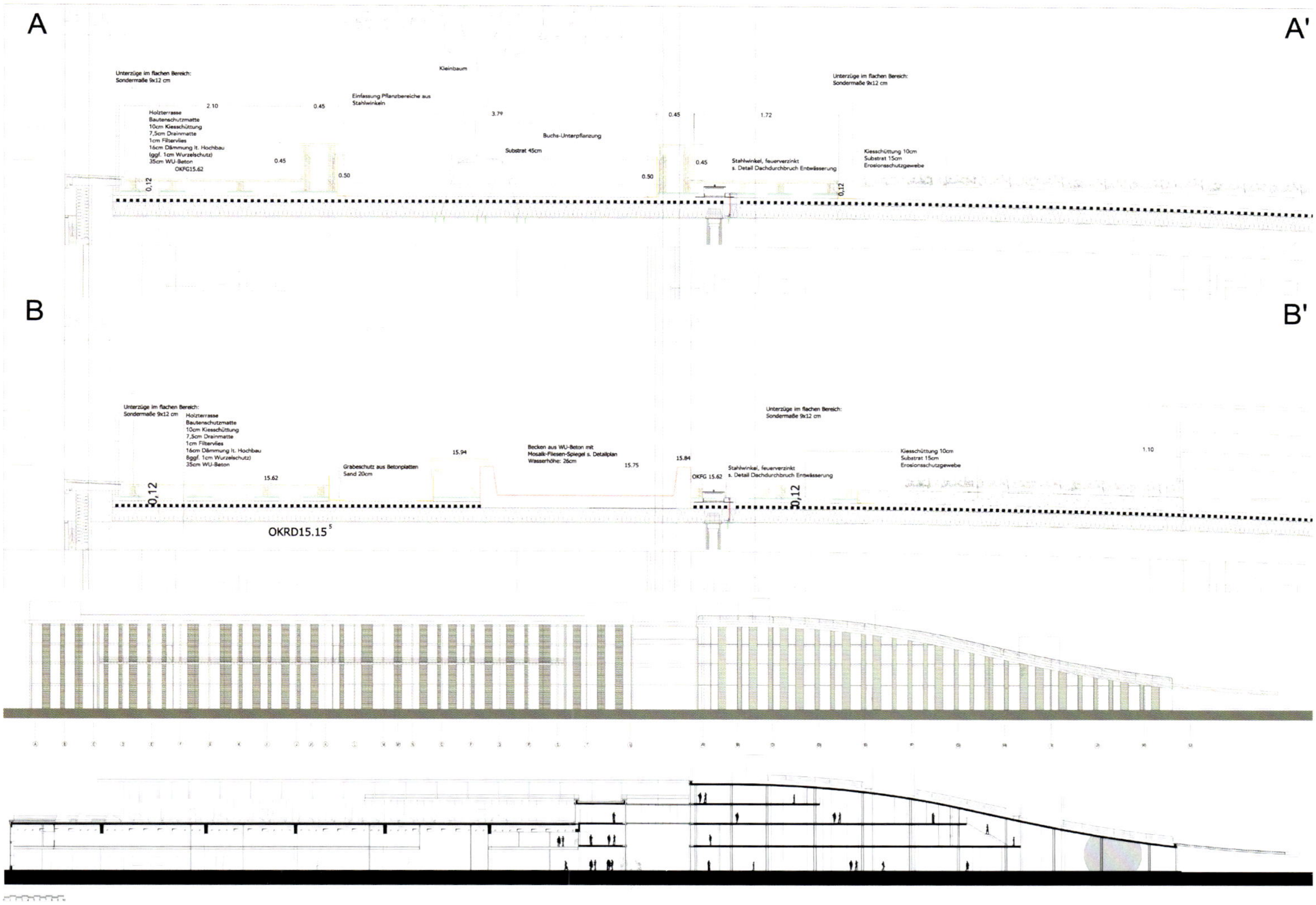
A
A'
Unterzüge im flachen Bereich:
Sondermaße 9x12 cm
Kleinbaum
Unterzüge im flachen Bereich:
Sondermaße 9x12 cm
Einfassung Pflanzbereiche aus
Stahlwinkeln
2.10
0.45
3.79
0.45
1.72
Holzterrasse
Bautenschutzmatte
10cm Kiesschüttung
7,5cm Drainmatte
1cm Filtervlies
16cm Dämmung lt. Hochbau
(ggf. 1cm Wurzelschutz)
35cm WU-Beton
OKFG15.62
Buchs-Unterpflanzung
Substrat 45cm
0.45
0.50
0.50
0.45
Stahlwinkel, feuerverzinkt
s. Detail Dachdurchbruch Entwässerung
Kiesschüttung 10cm
Substrat 15cm
Erosionsschutzgewebe
0,12
0,12
B
B'
Unterzüge im flachen Bereich:
Sondermaße 9x12 cm
Holzterrasse
Bautenschutzmatte
10cm Kiesschüttung
7,5cm Drainmatte
1cm Filtervlies
16cm Dämmung lt. Hochbau
8ggf. 1cm Wurzelschutz)
35cm WU-Beton
15.62
Grabeschutz aus Betonplatten
Sand 20cm
15.94
Becken aus WU-Beton mit
Mosaik-Fliesen-Spiegel s. Detailplan
Wasserhöhe: 26cm
15.75
15.84
OKFG 15.62
Unterzüge im flachen Bereich:
Sondermaße 9x12 cm
Stahlwinkel, feuerverzinkt
s. Detail Dachdurchbruch Entwässerung
Kiesschüttung 10cm
Substrat 15cm
Erosionsschutzgewebe
0,12
0,12
OKRD15.15 5

带有观景露台的行政区的屋顶由五个庭院打通，将日光引入空间深处。建筑内部有着巨大的流动空间，在自动扶梯井道围壁后隐藏有让人不易觉察的防火门，防火门只有火灾发生时才会关闭。上三层空间呈阶梯状布置为开放式办公区，并设置有厨房、储物柜、更衣室、充电站和座椅区。除了行政区的360个工作区，还有很多正式或非正式会晤的场所。一些独立的小空间可使人们不受打扰，全神贯注地工作，也可让人们有空间打私人电话或者私密会晤。

建筑立面上预装了一个独立的木框架，包括有三层阳光控制涂层的外部遮阳板、窗口以及保护大楼免受恶劣天气影响的水平木条。

工作区域各处都有连接线，装有最先进的性能达到250勒克司的9瓦的LED灯、充电器以及可根据需要随意设置的隔间和LCD屏幕。特别开发的能量储存装置可以在夜间或非用电高峰期通过屋顶的太阳能储存设备充电，这种能量储存装置能为整个工作站提供10个小时的电量。因此员工可以在屋顶、花园或者庭院等大楼内的任何一个地方办公。

厚度达340毫米的水泥预制板里设置有水管，这些水管回收并利用生产区产生的废热，在冬天可将这些废热充进集中供热系统以节约大量能源。SOLON公司计划在总部建造沼气池，以提供总部大楼所需的所有能源，从而实现碳中和。大楼内对应防火隔区还安装了5个局部通风系统，以尽量减少通风截面，使总部在绝大部分时间里都实现自然通风，只有在冬天才使用机械通风系统。

雨水被收集和储存在一个埋在地下的容积达到75000升的水罐里。这些储存的雨水被用来冲刷厕所和进行庭院的灌溉。多余的雨水流进景观园林的排水沟。一个装有蓄电池的双轴太阳能电池板给太阳能充电站提供能源，充电站可同时给10部运载员工和来访者的电动车充电。如果阳光明媚，这些电动车借助太阳能可运行1000公里。

Regelgeschoss EG_ Headquarter SOLON SE
1: 200

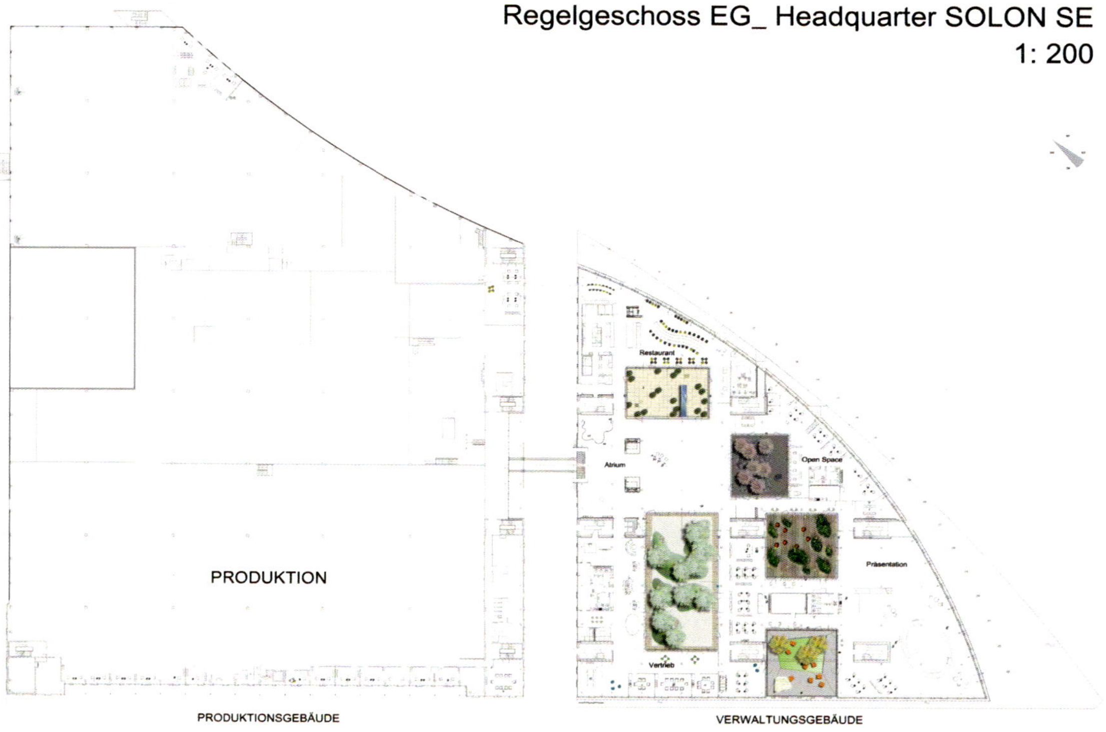

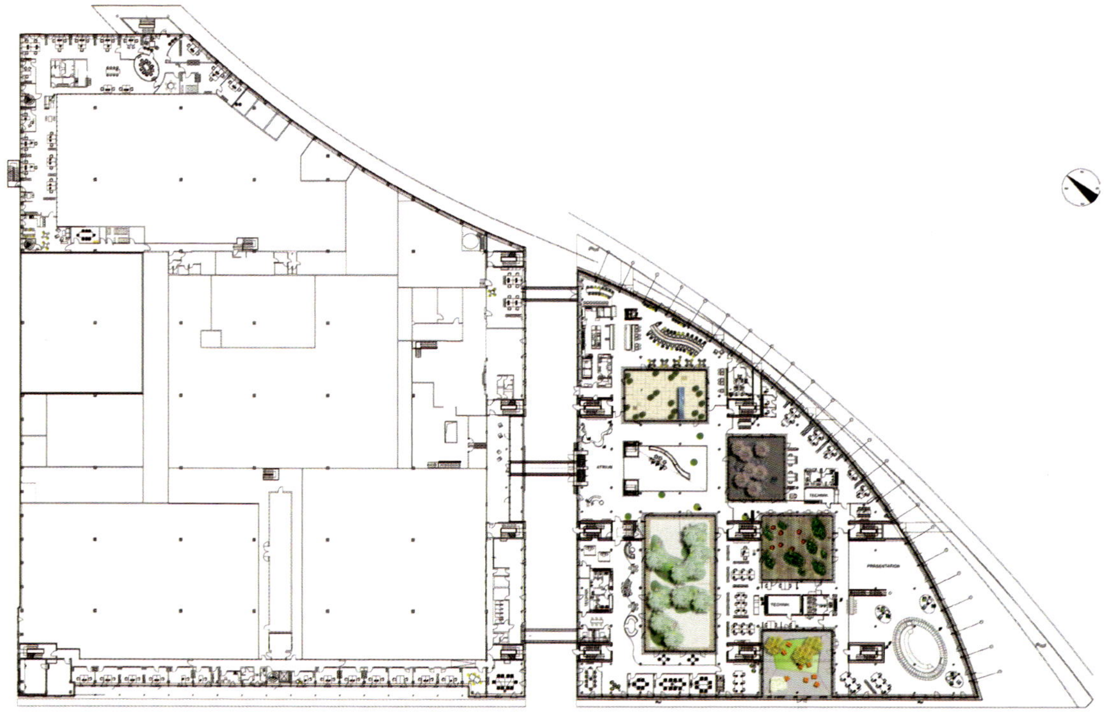

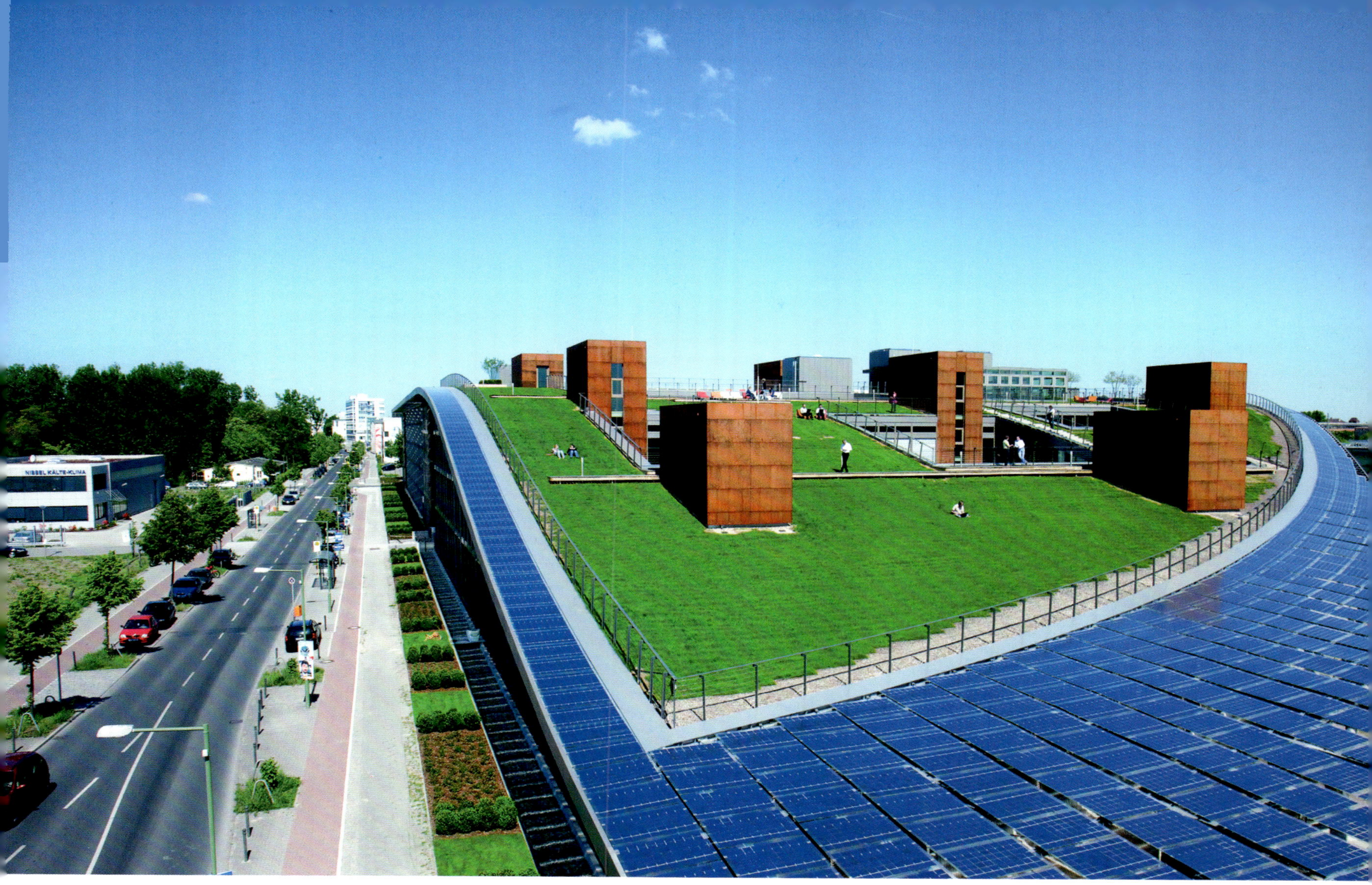

Schnitt Dachdurchbruch 1. Entwässerungslinie (+15.155)

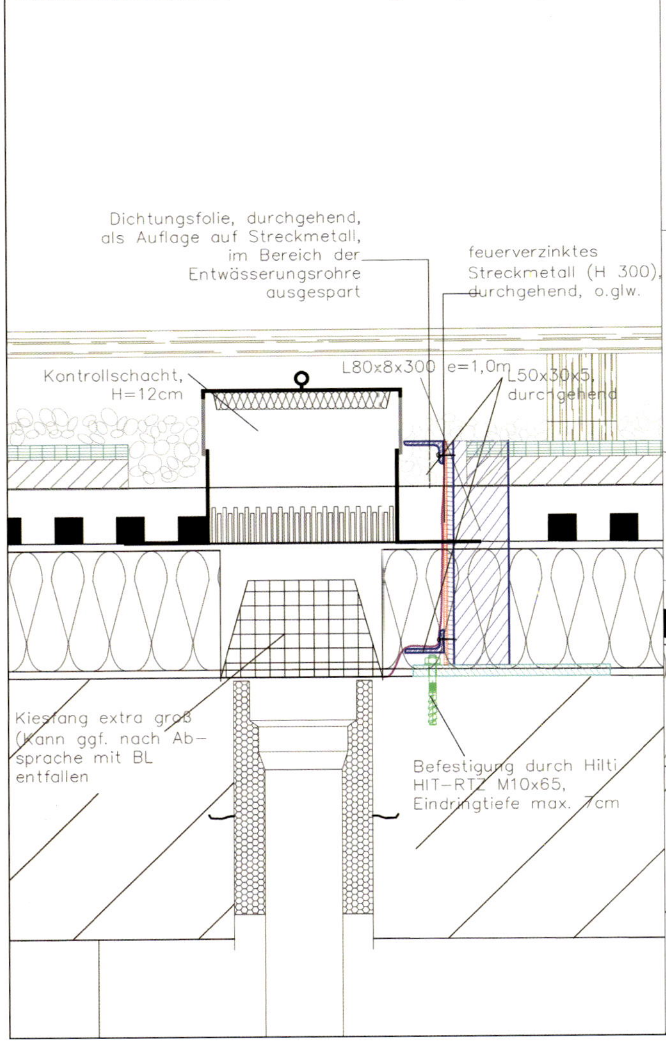

Aufsicht Flanschplatte

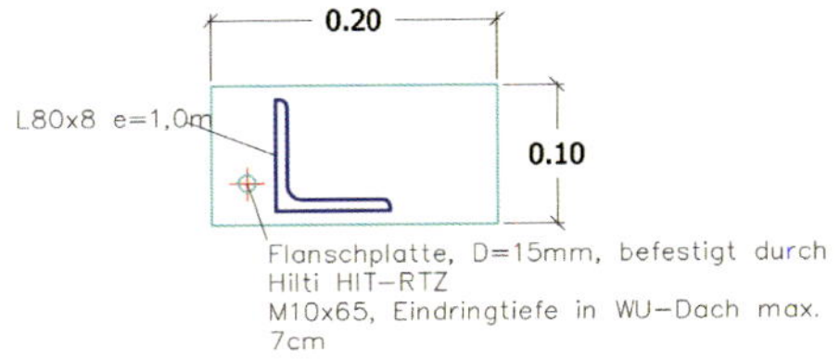

Schnitt Dachdurchbruch 2. + 3. Entwässerungslinie

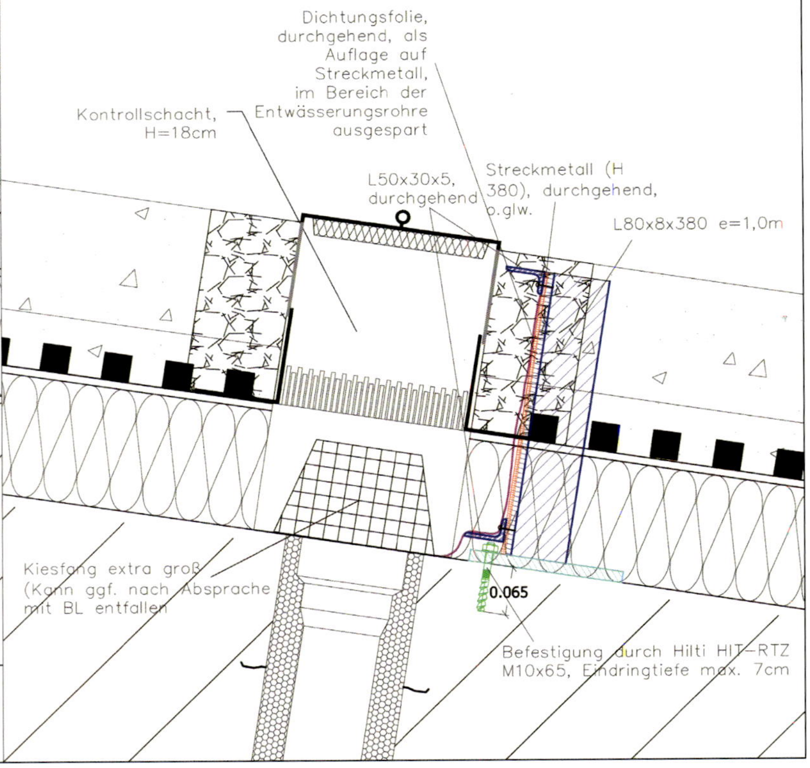

Podeste +9,80/+9,60

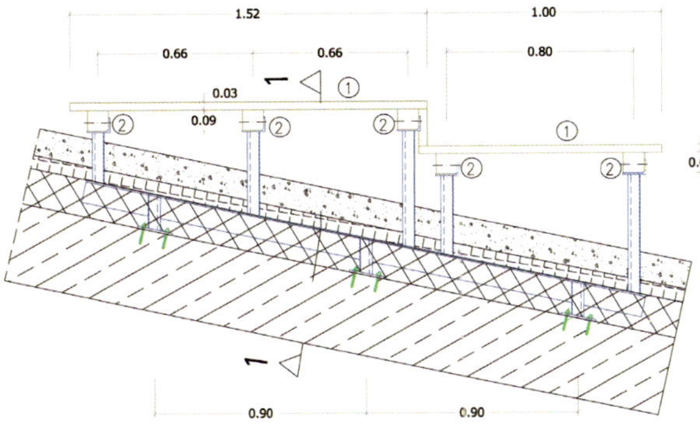

1. d=35mm LH D60 (Eiche)
2. b/d=90/90m² LH D40 (Eiche)
3. T100
4. L70x6
5. L100x65x7

Podeste +9,60

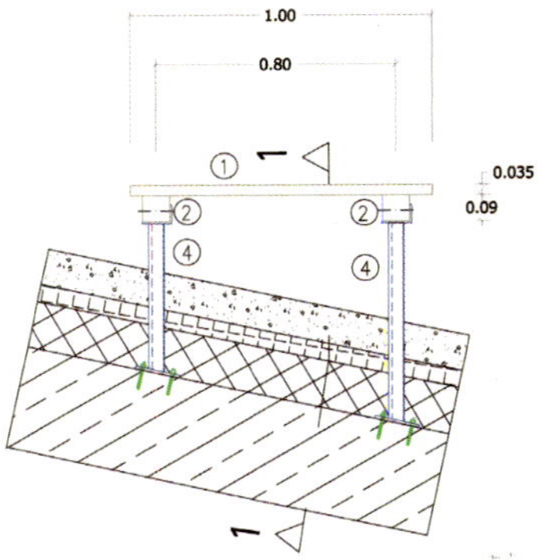

Schnitt 1 - 1

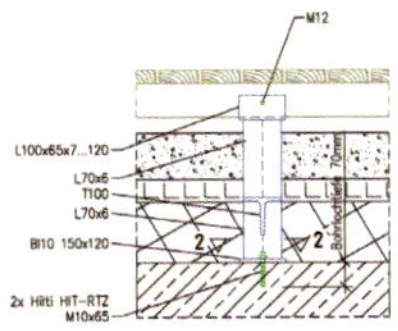

Podeste +14,34/+14,14
(Podeste +14,54/+14,34 analog)

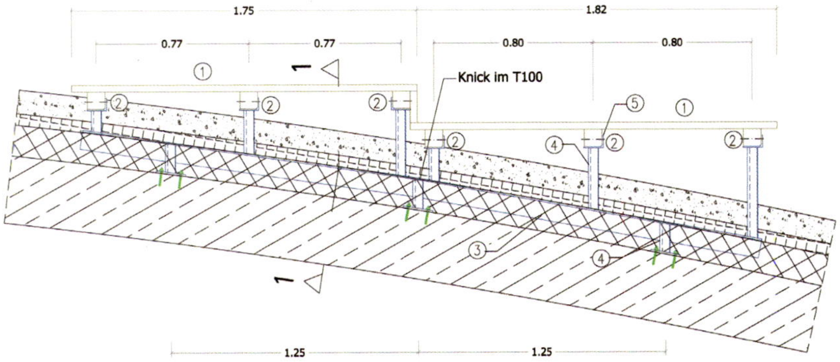

Schnitt 2 - 2

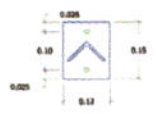

Dachrandbereiche < 5° geneigt
Abstand von OK WU-Attika: 10 cm

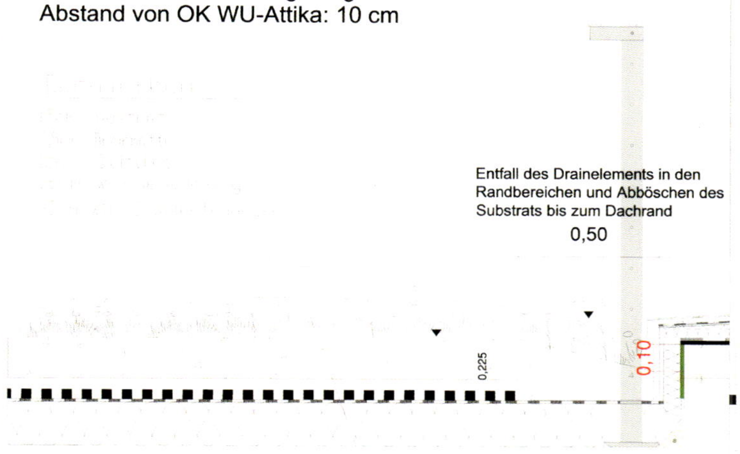

Dachrandbereiche > 5° geneigt
Abstand von OK WU-Attika: 5 cm

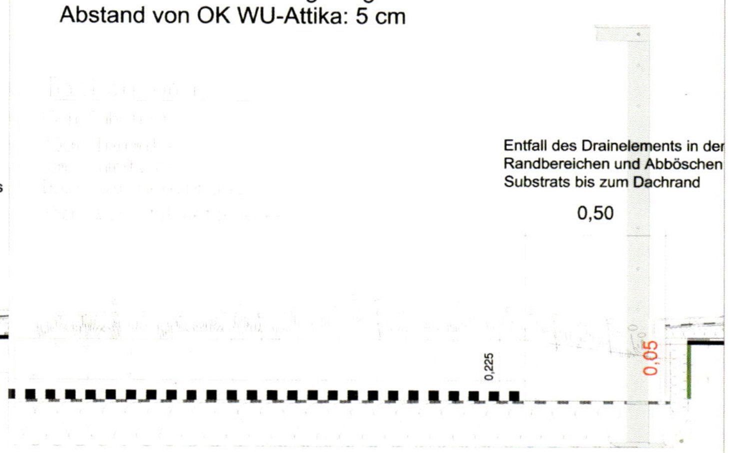

unterster Dachrandbereich Achse I-L
0,5 m Kiesrand, da keine weiteren Wassersammelprofile vorhanden -

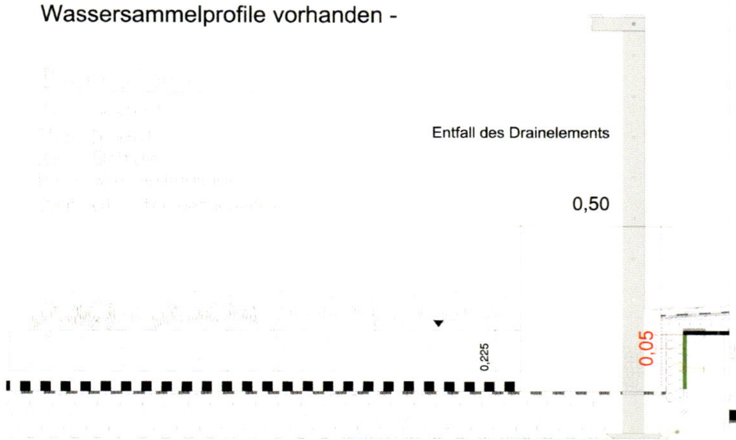

Hofrandbereiche sowie Randbereich Voigt-Terrasse :
Ausbildung einer offenen, kiesgefüllten Rinne durch Trennwinkel an Geländerstützen angelehnt

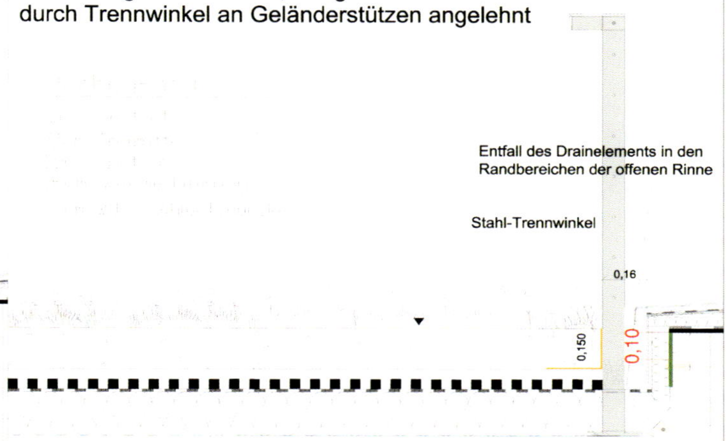

Hofattika im Bereich des Treppenlaufs:
Ausbildung einer offenen, kiesgefüllten Rinne abgetrennt durch Betonkantensteine des Treppenlaufs

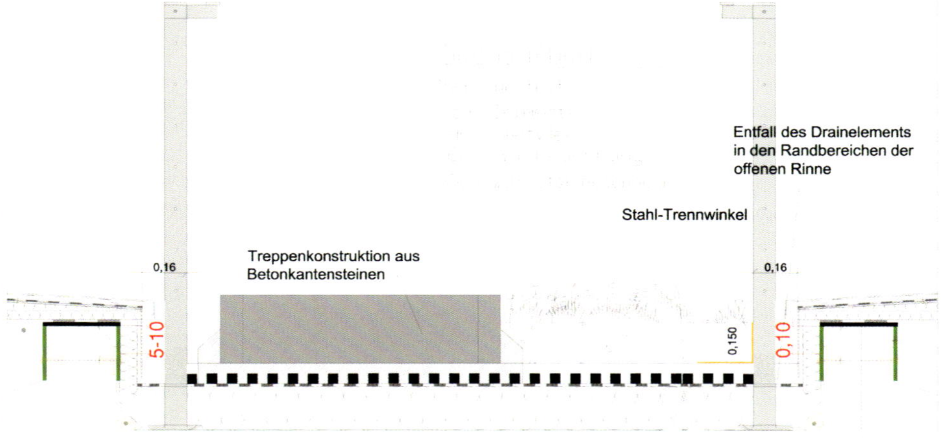

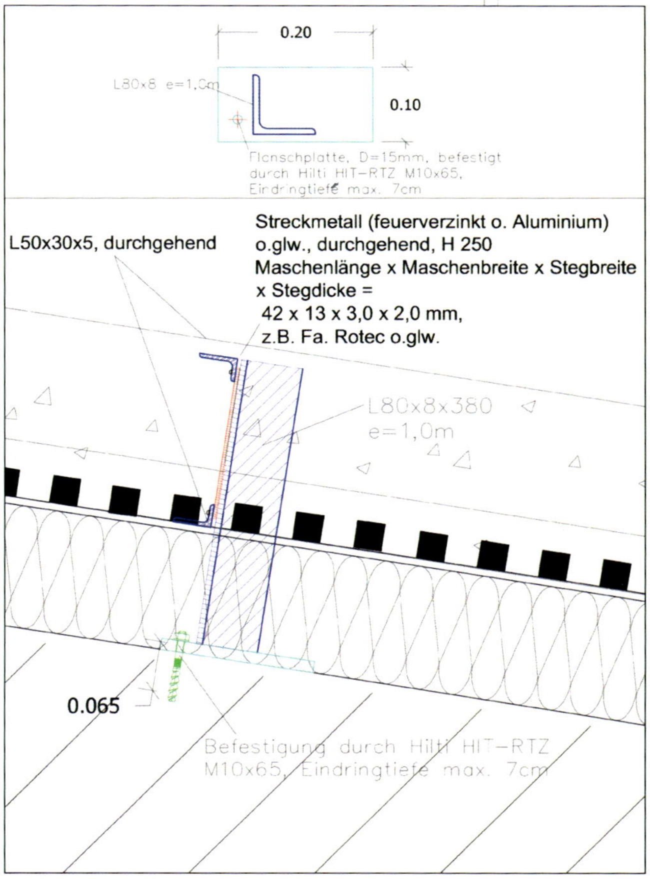

Rivergate

Rivergate, the office centre on the Danube, placed in immediate vicinity of the Millennium Tower, constitutes a strong accent within Waterfront Vienna.

Two buildings of different heights opening towards the Danube form a generously shaped atrium. A public walkway through the project helps to mark clearly the access to the building via the atrium. The free-form shell has become characteristic for the skyline on the Danube, it enhances the flexibility of use by varying building sizes and allows to react appropriately to building regulations.

LOCATION\Handelskai 92, 1200 Vienna, Austria

ARCHITECT
Auer+Weber+Assoziierte, Stuttgart
GENERAL CONTRACTOR
Neumann+Partner, Vienna
STRUCTURAL PAUSER
ZT-GmbH
MECHANICAL
ZFG - Projektges. m.b.H
ELECTRONICAL
ZFG - Projektges. m.b.H
CONSTRUCTION MANAGEMENT
Pro Projekt, Vienna
AWARDS
DIVA-Award 2010, Green Building Zertifikat der EU, Für Energiebewusstes Bauen, LEED zertifikat des USGBC mit Gold Status 2010
AREA
80000 m^2
PHOTOGRAPHER
Rupert Steiner

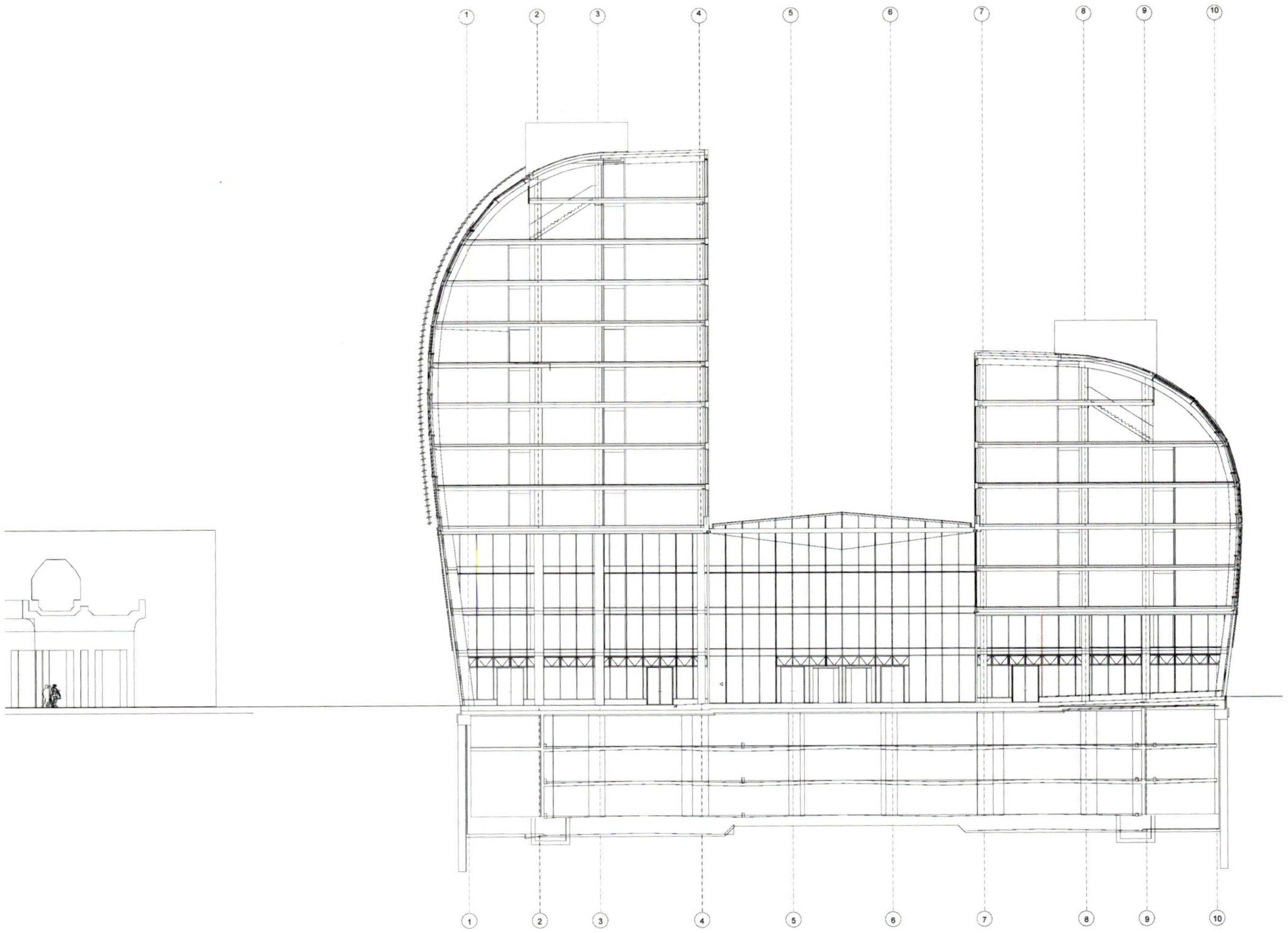
1
2
3
4
5
6
7
8
9
10

WWW.RIVERGATE.AT

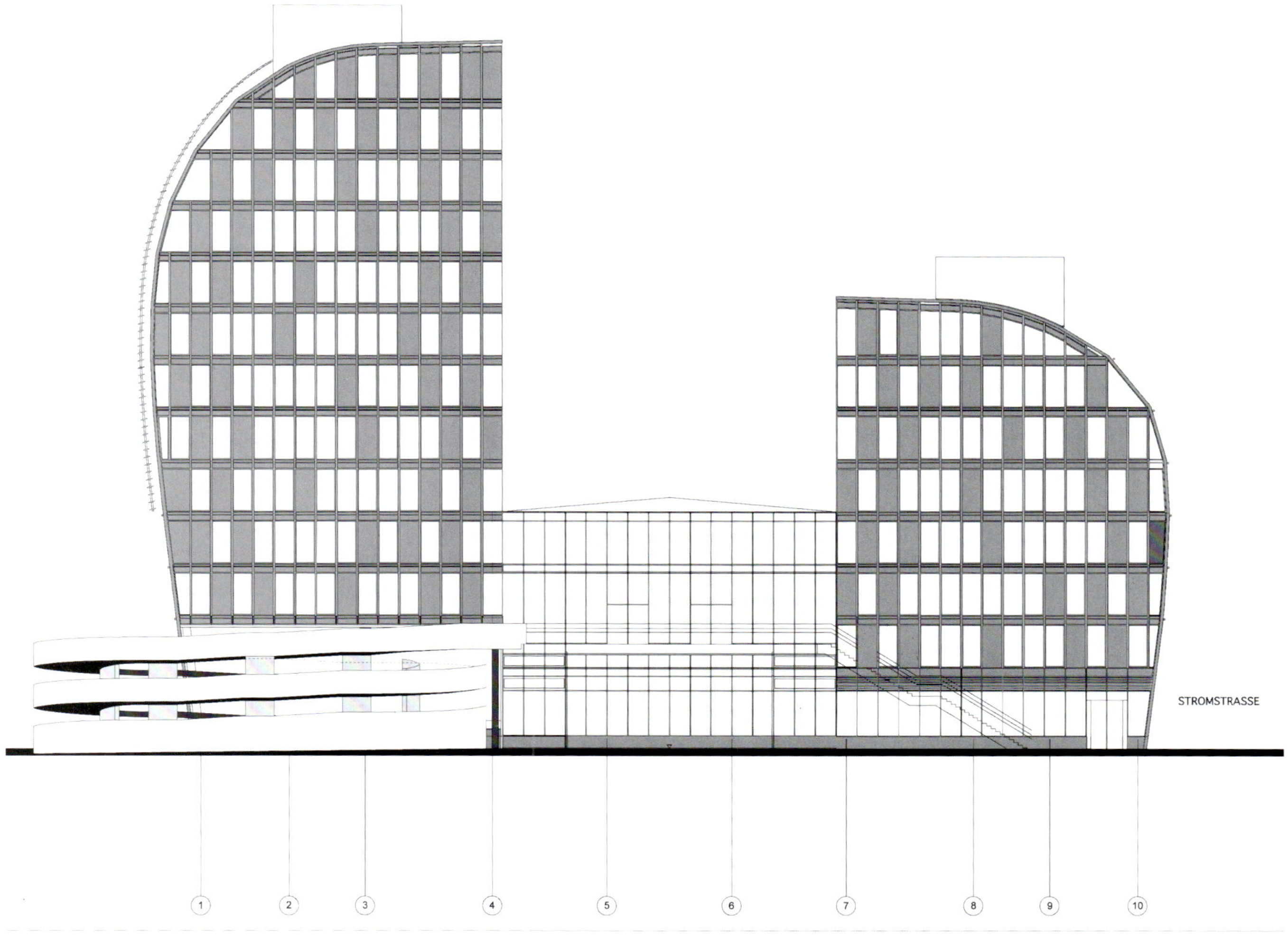
STROMSTRASSE
1
2
3
4
5
6
7
8
9
10

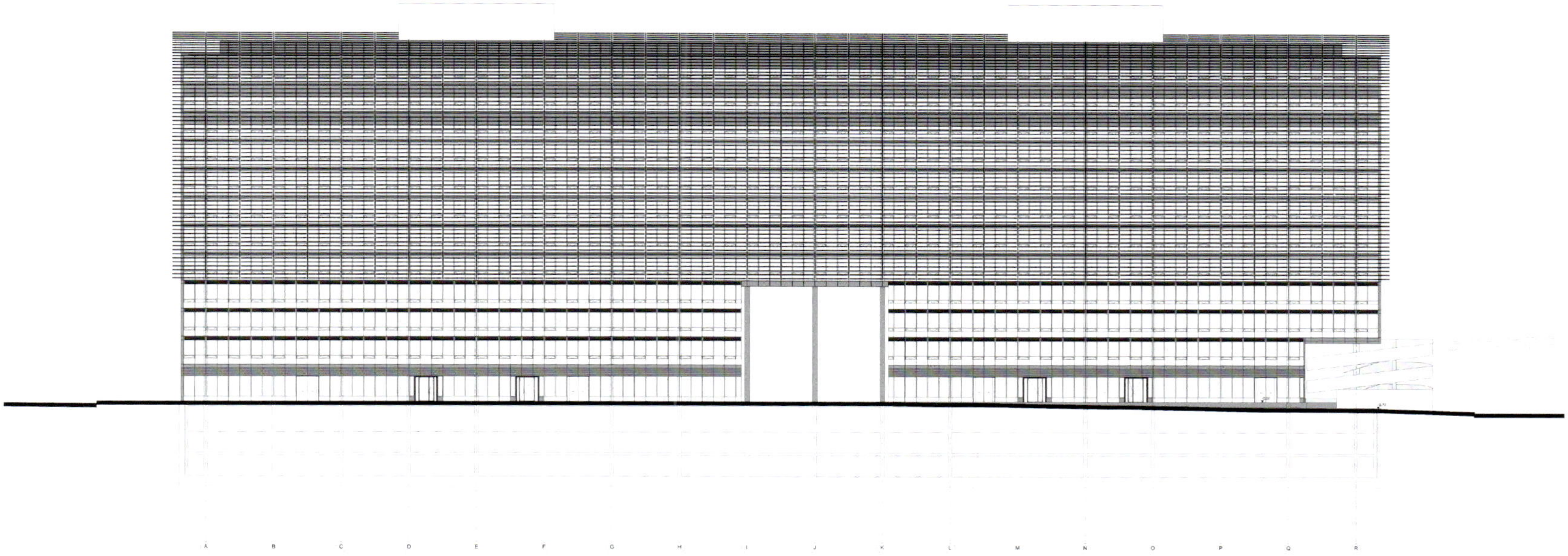

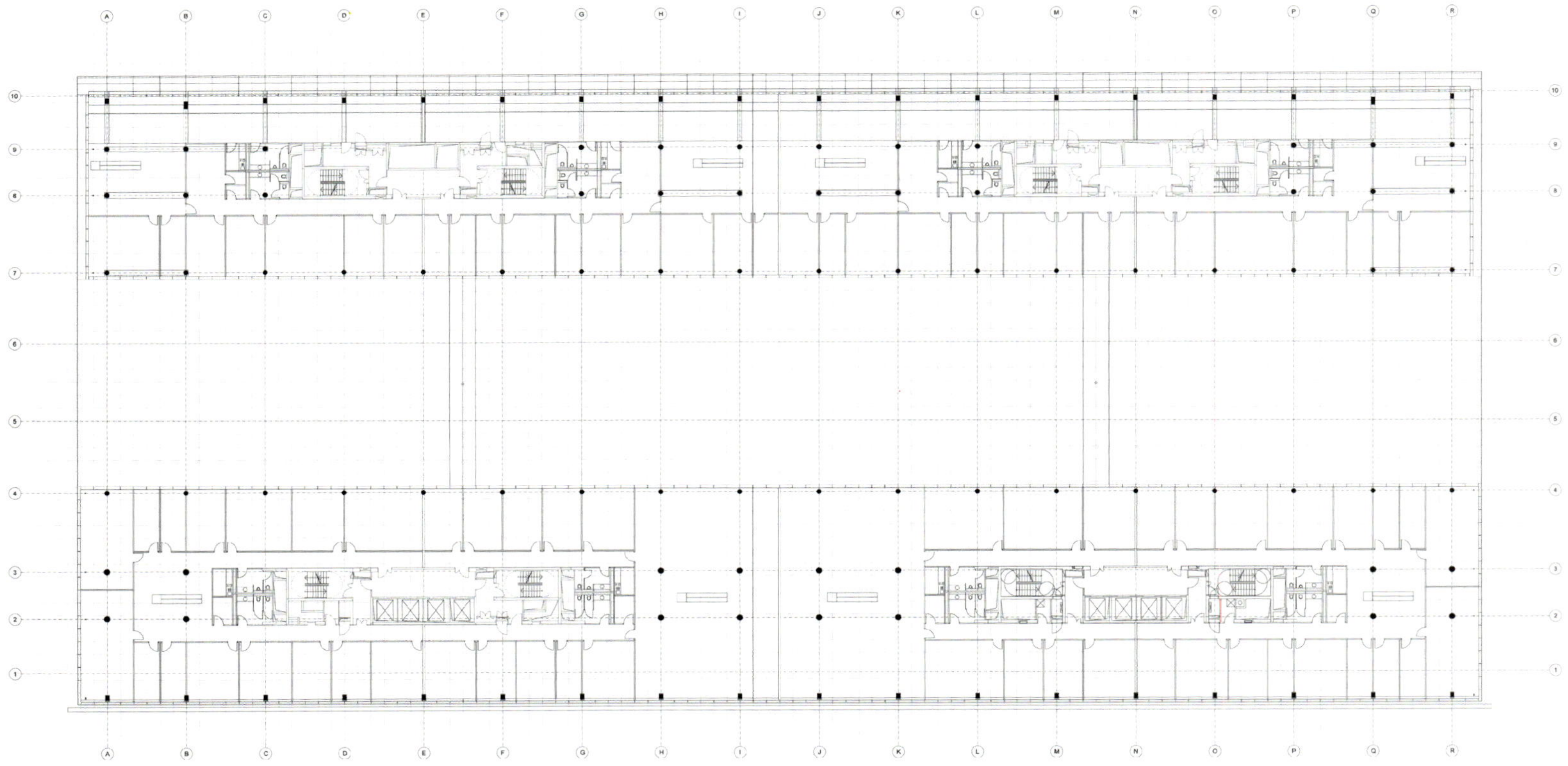

UPPER FLOOR

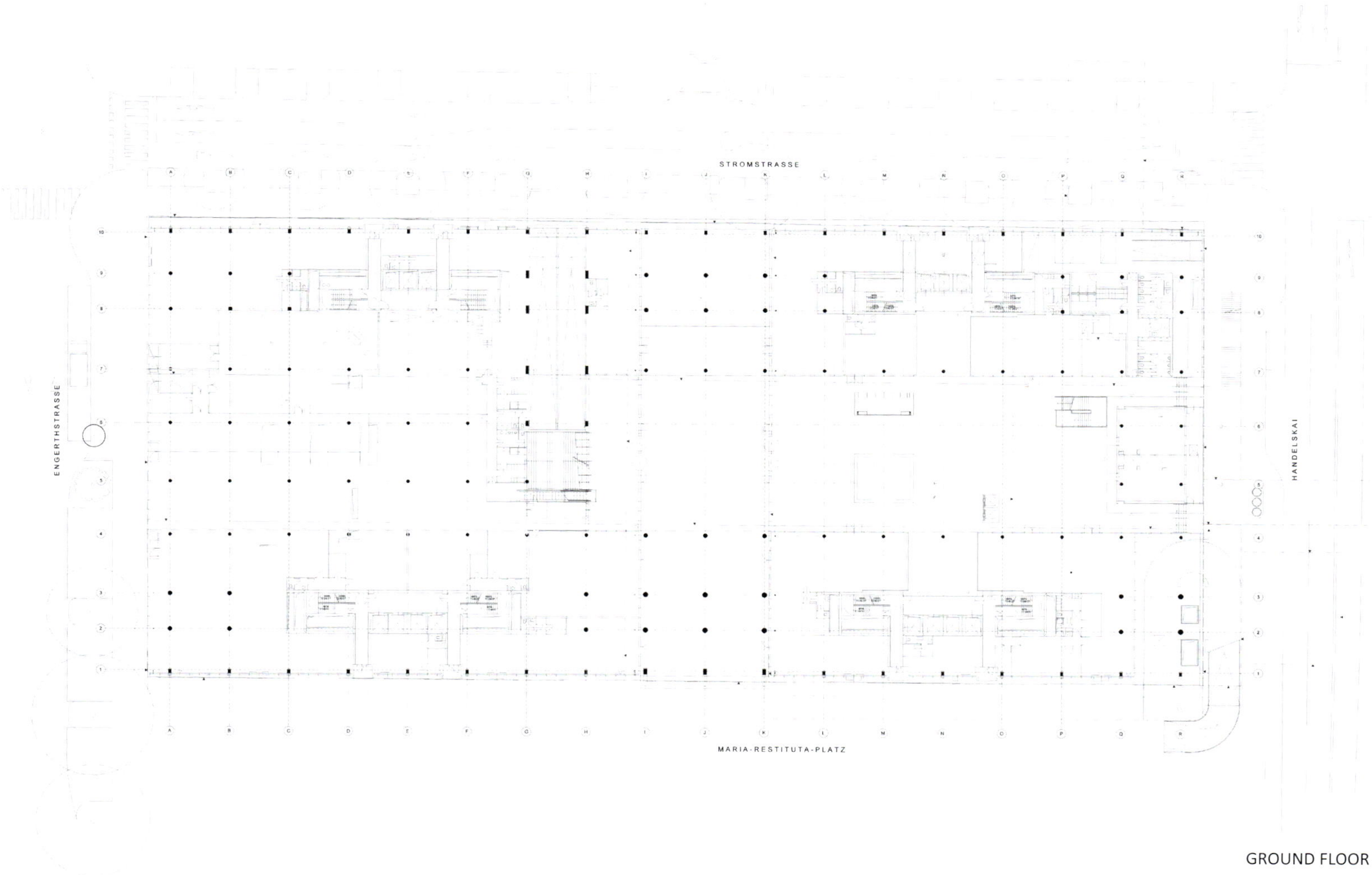

GROUND FLOOR

Rivergate 办公中心耸立在多瑙河畔，临近千禧塔，在滨水地区拥有不可替代的地位。

两个不同高度的大楼形成了一个巨大的天井，朝向多瑙河。在中心外面的人行道上，很容易找到经由天井进入大楼的入口。建筑的外观形成了多瑙河畔一道独特的风景线，使用不同的建筑规格加强了建筑的灵活性，同时也遵守了相应的建筑规程。

restaurant · cafe · bar
Canteen

D2
2.UG

The Water Centre

Located on a former brownfield site, this 17,000 m^2 LEED Gold workplace for 800 employees spearheads the City of Calgary's recent mandate that all their new buildings achieve minimum LEED Silver rating.

Inspired by the movement and translucency of water, the building also represents the first phase of the site's redevelopment to a satellite civic office centre. Sustainable design features will include a reduction in energy and water consumption while creating a healthy, dynamic working environment, enhancing the quality of life for staff. Using an integrated design process, they arrived at a narrow floor plate to optimize daylight in all workspaces, while enhancing collaboration in a large south-facing atrium.

Usually, industrial areas tend not to be pedestrian-friendly, but The Water Center has an LRT station nearby. The building sits on the edge of the site allowing space for a large garden of native plants, which will be an amenity for both the city workers and the residents of nearby neighborhoods. The building also boasts fantastic views of the downtown core, and the bold shape of the building can be seen from many vantage points around the downtown area.

LOCATION\Calgary, Alberta, Canada

ARCHITECT
Vivian Manasc, Derek Heslop
FIRM
Manasc Isaac
STRUCTURAL CONSULTANT
Read Jones Christofferson Calgary
MECHANICAL AND ELECTRICAL CONSULTANT
Keen Engineering (now Stantec)
CIVIL CONSULTANT
Urban Systems
LANDSCAPE CONSULTANT
Carlyle & Associates, Harris and Harris Landscape Architects
COSTING CONSULTANT
Spiegel Skillen & Associates
ACOUSTICAL
ACI Acoustical Consultants
AREA
17000 m^2

YIELD
ON
SOLID
GREEN

WATER

CENTRE

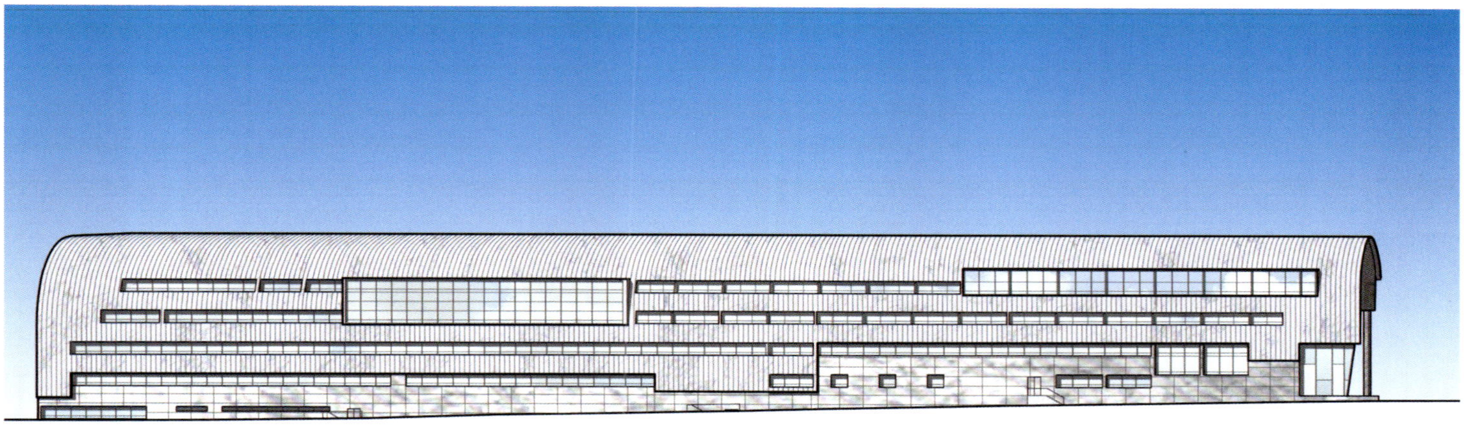

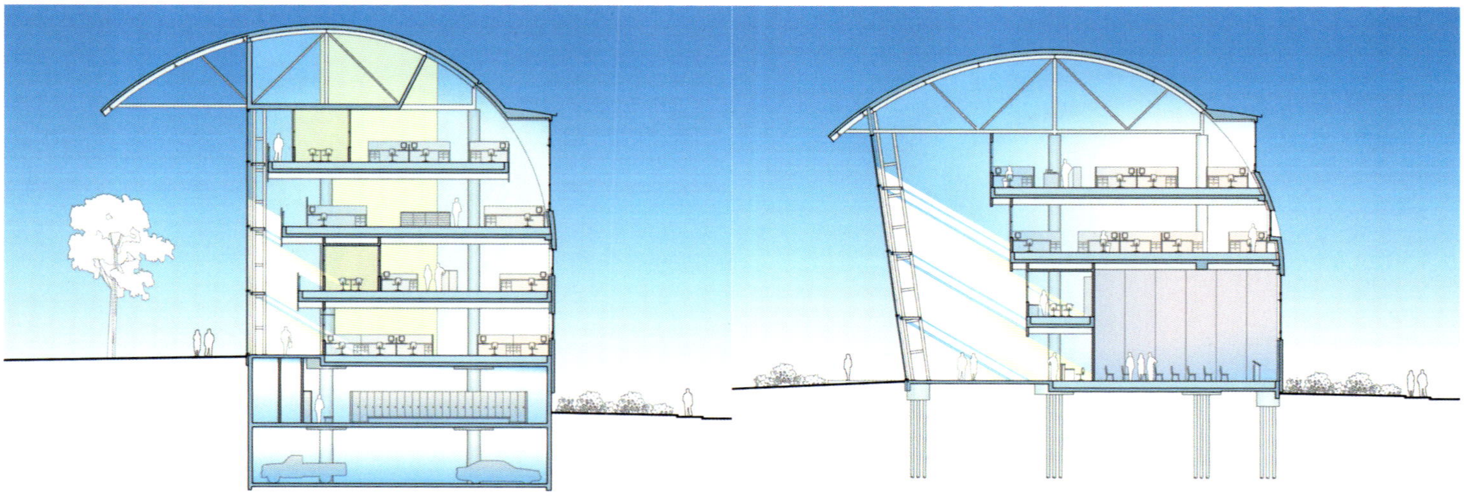

本案位于一个以前荒置的场址，这块 17000 平方米的土地是一个拥有 800 个员工的工作场所。科尔加里市最近的法令规定所有新建筑物必须取得绿色建筑评估体系 LEED 认证的最低等级，即银级，而水中心因获得 LEED 认证中的金级而成为环保先锋。

本案的设计灵感来自于水的运动和水的半透明性，这座建筑物也代表了荒地再开发的第一阶段：一个卫星市政办公中心。可持续的设计特点包括降低能耗与用水量，创建一个健康的、有活力的工作环境和提高员工的生活质量。设计使用集成设计程序，用一块细长的楼面板使所有的工作空间最大限度地使用日光，并由一个大的朝南的中庭把这些空间连为一体。

通常，工业区往往交通不便，但水中心附近却有一个轻轨车站。这栋建筑的旁边还有空间可以建一个很大的长着原生植物的花园，将会是城市工人和附近居民休闲娱乐的宜人场所。这座建筑物能将市中心的美景尽收眼底，其奇异的建筑造型也可以从市区中心多个有利地点欣赏到。

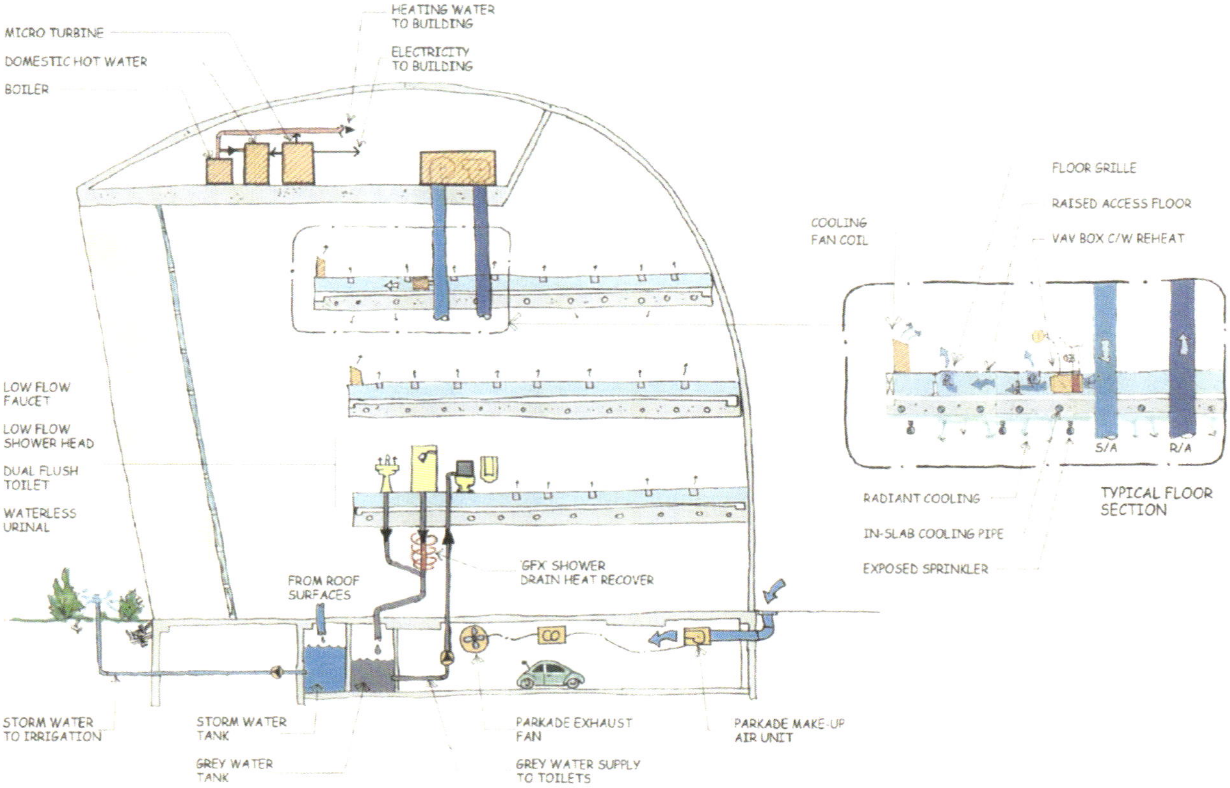
MICRO TURBINE
DOMESTIC HOT WATER
BOILER
HEATING WATER TO BUILDING
ELECTRICITY TO BUILDING
FLOOR GRILLE
RAISED ACCESS FLOOR
COOLING FAN COIL
VAV BOX C/W REHEAT
LOW FLOW FAUCET
LOW FLOW SHOWER HEAD
DUAL FLUSH TOILET
WATERLESS URINAL
S/A
R/A
RADIANT COOLING
TYPICAL FLOOR SECTION
IN-SLAB COOLING PIPE
EXPOSED SPRINKLER
GFX SHOWER DRAIN HEAT RECOVER
FROM ROOF SURFACES
STORM WATER TO IRRIGATION
STORM WATER TANK
GREY WATER TANK
PARKADE EXHAUST FAN
GREY WATER SUPPLY TO TOILETS
PARKADE MAKE-UP AIR UNIT

EXIT
EXIT

Torre Arrau

The building's layout is divided into two integrated areas around a core, which receives the common spaces of accesses, elevators and services. Located at the building's ends, and open towards the facades, are the emergency stairs, which are treated as balconies, thus enabling the use of the building's exterior areas and incorporating natural light to its central hall.

Stairs in turn are projected as separate volumes that stand out as distinctive components, and generate interstices that allow natural light to the center of the building.

Energy efficiency has been incorporated as a base concept for the project, through the use of last generation air conditioning systems, along with a façade design based on a floor to floor enclosure of Low-e thermo panes and an angled reflective skin that induces ventilation. The distancing from the façade as well as the separation between the crystal layers of the skin, allow the inner façade's ventilation, contributing to improve the building's energy performance.

The proposal's double façade generates a perception of the workspace that expands its boundaries beyond the building, while containing it through the effect of the transparent skin veil, becoming a permeable space which enables continuity with the exterior.

The project envisages a façade vertical continuity break every four to five floors, through the manipulation of the distance between the perimeter axes and the slab edge, in order to avoid ventilation's convection effect.

The building's insertion in the public space recognizes its particular location in the urban fabric that relates the BOULEVARD CENTRAL with PRESIDENTE RIESCO Avenue, proposing a passable public wooden Deck, accentuated with a water mirror that gives the ground a connecting role between the interior of the complex and the rest of the urban area. This solution seeks to transform the buildings' site in a transit space as well as a public square.

The project image, that defies the anonymity resulting from the indiscriminate use of curtain wall solutions in the monolithic massing of surrounding buildings, is achieved through an active volume that plays with scales, façade movements and a permeable enclosure, reflecting on its volumetric image the concepts of energy control, transparency and inside-outside continuity, by the use of a technological and contemporary language.

ARCHITECT
Juan Sabbagh P.-Mariana Sabbagh P.- Juan Pedro Sabbagh B.-Felipe Sabbagh B.- Mario Zamorano V.

FIRM
Sabbagh Arquitectos

ARCHITECT COLABORATOR
Miguel Ángel Herrera M.

TECHINCAL COLABORATOR
BMV- Mónica Pérez- María Teresa Leighton- Gormaz y Centeno- Energy Arq

AREA
44880 m^2

LOCATION AV. Presidente Riesco 556, Las Condes, Santiago, Chile

Plantas libres
Arriendo
371 00 00

该大楼的核心区域是一个集入口、电梯空间和服务空间于一体的公共区域，围绕这个核心区域，建筑被分为相对完整的两个单元。楼层两端没有封闭，开放的空间朝向立面敞开，兼做安全出口的疏散楼梯和露台。这样的布局既利用了建筑的外部空间，也使自然光能照进中央大厅。

设计使楼梯成为独立的空间，为该建筑增添一丝独特性，所形成的立面间隙也使大楼中心光线充足而明亮。

该工程设计的基本理念是高效率地利用能源。楼内使用的都是新一代节能空调。立面由低辐射隔热玻璃板形成围护结构，在围护结构上覆上成一定角度的反光幕墙膜材料，这种设计使整座建筑通风良好。幕墙材料与立面之间留有一定空隙，其各层次间成分离状态，使内部通风良好。这些设计理念和措施大大提高了建筑本身的节能性能。

设计方案中采用了双层立面，使用了具有透明效果的表皮膜，形成了渗透空间。这些措施让工作空间得以延展，使其超越了建筑本身设定的边界，与外界保持某种形式的连续，从而形成整体性的效果。工程还有一个显著特点就是立面在竖直方向上是连续完整的，每隔四到五层的楼层距离才会有连接的痕迹。在连接处通过对周边轴线和面板之间距离的调节来避免空气的强对流效应。

该建筑在整个城市结构中处于非常特殊的位置，嵌入中央林荫大道和 PRESIDENTE RIESCO 大道之间的公共区域，两条大道通过此处连接。计划在项目地基处建造一条可供行人通行的木质栈道，并重点打造一面“水镜”以使该大楼的内部景观与周围的城市环境互相交融，浑然一体。这种设计方案力图改造该楼盘作为过路区和公共广场的形象和地位。设计中绝对不会随随便便使用幕墙方案，让自己淹没在周围林立的建筑中，而是在规模与立面的形态变化、外围结构的渗透等方面做足文章。最终呈现的整体形象为：通过高科技的手段和当代建筑语汇的结合使用，节能、透明，楼内外浑然一体；既与周围环境承接，也显得特色鲜明。

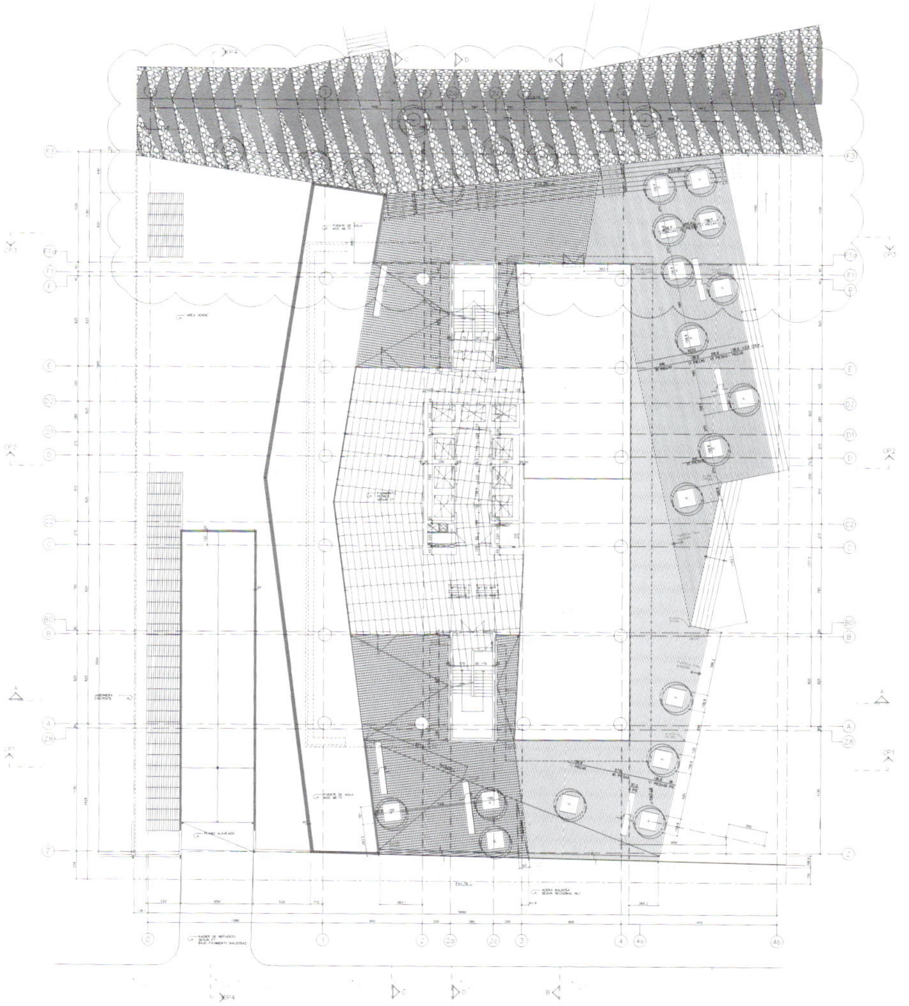

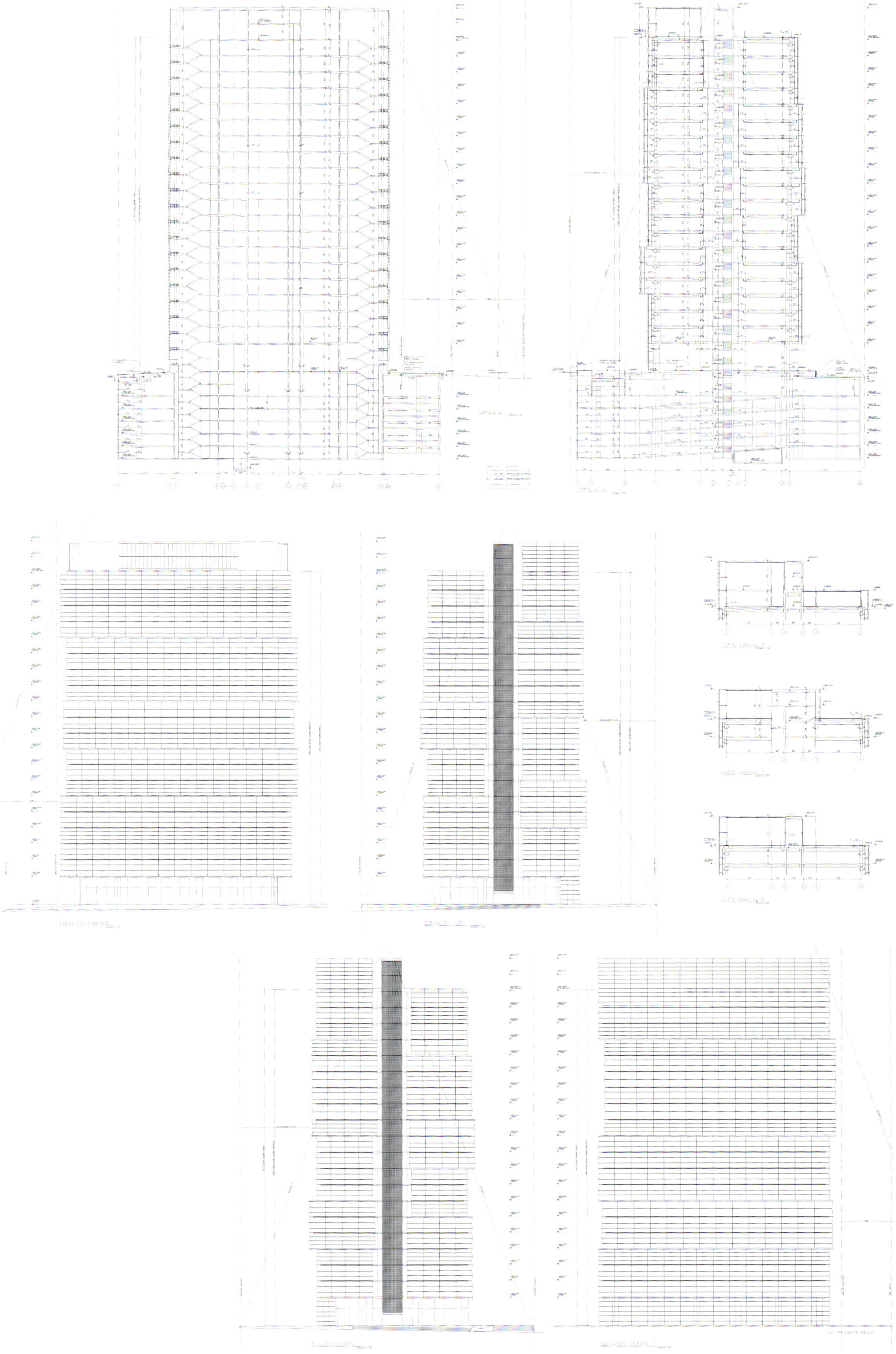
ELEVACION PONIENTE
ELEVACION SUR
ELEVACION ORIENTE
AV. PRESIDENTE RIESCO

5561

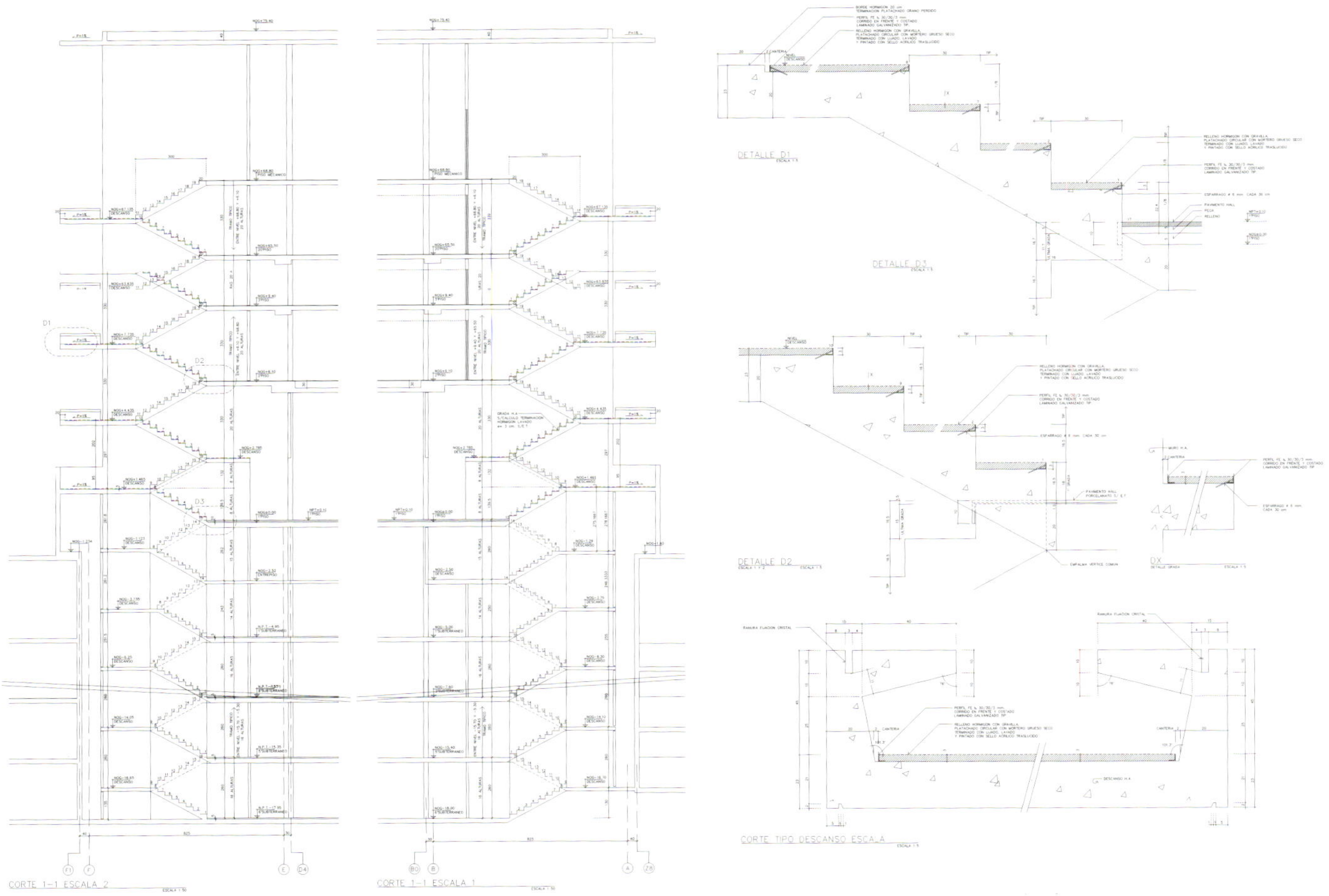
DETALLE D1
DETALLE D3
DETALLE D2
CORTE TIPO DESCANSO ESCALA
CORTE 1-1 ESCALA 2
CORTE 1-1 ESCALA 1

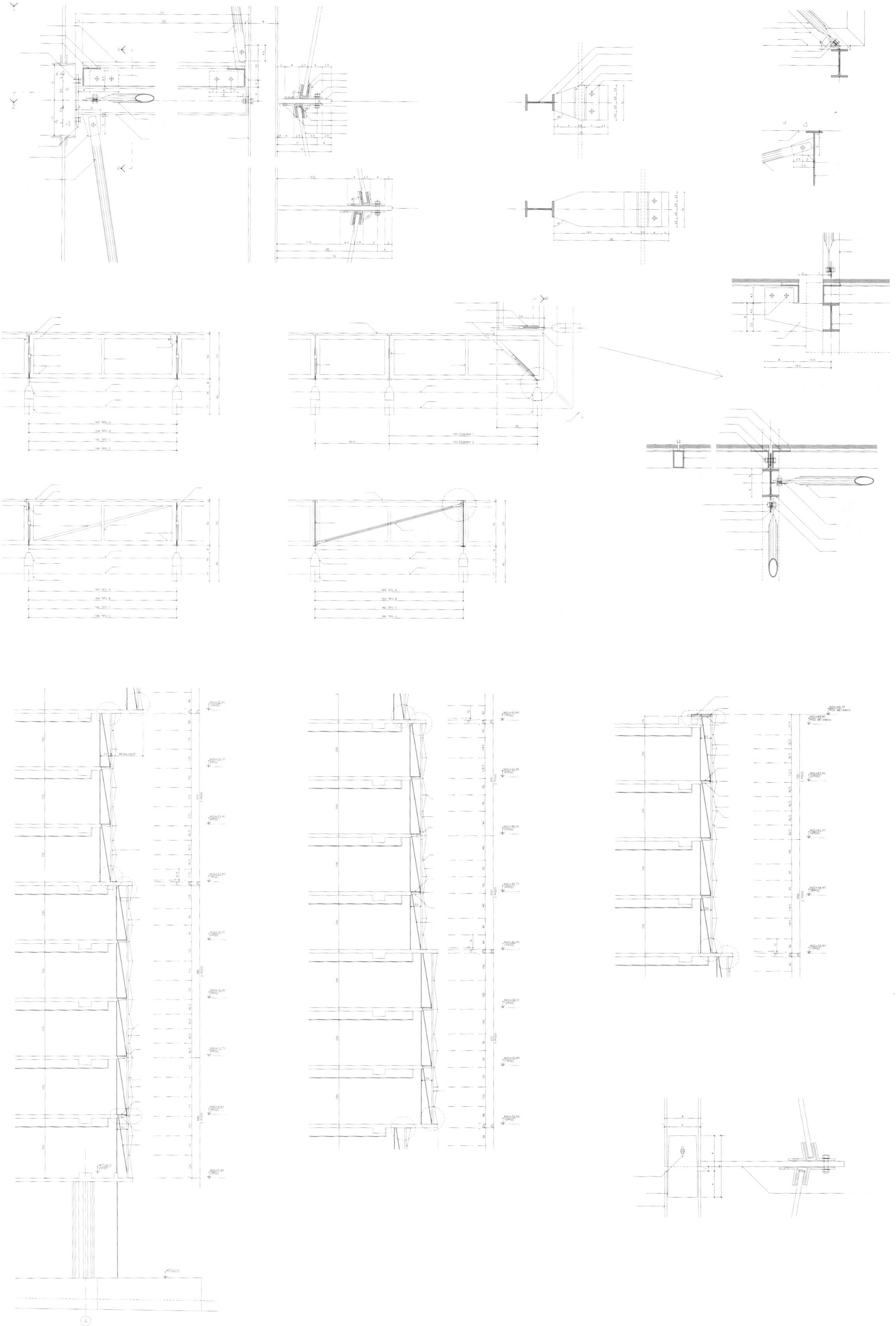

SKIDMORE, OWINGS & MERRILL LLP

SKIDMORE, OWINGS & MERRILL LLP

Skidmore, Owings & Merrill LLP (SOM) is one of the leading architecture, interior design, engineering, and urban-planning firms in the world, with a 75-year reputation for design excellence and a portfolio that includes some of the most important architectural accomplishments of the 20th and 21st centuries. Since its inception, SOM has been a leader in the research and development of specialized technologies, new processes and innovative ideas, many of which have had a palpable and lasting impact on the design profession and the physical environment. The firm's longstanding leadership in design and building technology has been honored with more than 1,600 awards for quality, innovation, and management. The American Institute of Architects has recognized SOM twice with its highest honor, the Architecture Firm Award—in 1962 and again in 1996. The firm maintains offices in New York, Chicago, San Francisco, Los Angeles, Washington, D.C., London, Hong Kong, Shanghai, Mumbai, and Abu Dhabi.

MANASC ISAAC

Building for a Climate of Change: Manasc Isaac is an innovative, award winning architectural team. Their projects reflect the culture of the unique regions, communities, and organizations with whom they collaborate. They are among Canada's leaders in the design of high-performance new and renovated buildings to achieve net-zero communities and buildings. Dedicated to a client-centered and workshop based planning, design and construction approach to architecture, Manasc Isaac has a reputation for architecturally stunning and technically outstanding buildings.

Their firm has been recognized with a number of awards, including the first-ever Brilliant Building Award, the Governor General's Award in Architecture (the profession's highest national commendation) and a Royal Architectural Institute of Canada Innovation Award.

www.manascisaac.com

ARQUITECTONICA

Arquitectonica is a design architecture, master planning, interior design and landscape firm founded 1977 in Miami as an experimental studio. Led by Bernardo Fort-Brescia (FAIA) and Laurinda Spear (FAIA, ASLA) the practice has evolved into an international studio of design professionals located across several continents and eleven offices. Arquitectonica combines the creative spirit of the founding Principals with the efficient delivery and reliability of larger corporate architectural firms.

Arquitectonica's creative reputation is widely recognized through its large and growing portfolio of built-projects and multiple-design awards; with more than eighty built projects across thirty international cities and over one-hundred and twenty awards for building design through American Institute of Architects (AIA), Progressive Architecture and other high-profile award sponsors, Arquitectonica proves a track record of service delivery, and designing commercially successful buildings recognized through the value of award winning design.

KRUECK & SEXTON ARCHITECTS

Since its emergence to national attention with a Steel and Glass House in 1981, Krueck & Sexton has developed a design approach characterized by devotion to client needs, impeccable detailing and a dedication to craft. The firm's work honors and reveals the designers, the craftspeople and the inhabitants, which has provided a foundation for this small firm to excel at producing diverse and large scale work.

As founding partners of Krueck & Sexton, Ron Krueck and Mark Sexton play complimentary roles. Ron is responsible for the firm's architectural vision and body of work. Mark manages all of the firms work. Together they design in partnership, approaching each project as unique with its own set of opportunities.

Both principals graduated from the College of Architecture at the Illinois Institute of Technology, after which Ron went on to study painting at the Art Institute of Chicago.

Ron and Mark's belief that exceptional architecture can and does make a difference to people's lives is a hallmark of the firm's work, and adds value not only to the profession of architecture but to our society as a whole.

Krueck+Sexton's work has been honored with 23 design awards from the American Institute of Architects. Nine of their completed projects have been recognized as being exemplary in energy and environmental design performance through the LEED certification program, one of which received the prestigious AIA/COTE Top Ten award. The firm's environmental commitment is further exemplified in the future of the firm which includes a net zero 2030 US Federal Office Building for the GSA Design Excellence program.

TAKASHI YAMAGUCHI

1953 Born in Kyoto. Graduated Kyoto University. Worked in Tadao Ando Architect & Associates.

1988 Founding member of ARX, between New York, Geneve, Berlin, and Lisbon, and international research group for theory and practice of architecture, with staffs of Eisenman Architects and Daniel Libeskind's office, and theorists in Columbia University.

1996 Founded Takashi Yamaguchi & Associates

2001 Senior Scholar of Waseda University / IMDC

2005 Visiting Professor, Osaka University of Arts (Japan)

2009 Visiting Professor of Columbia University, GSAPP (US)

Visiting Scholar of Harvard University, GSD (US)

2010 Elite Member of the Supporters of Romania Research: THE ACADEMY of ROMANIAN SCIENTISTS (Romania)

Present Professor of Osaka Sangyo University (Japan)

AWARD

2009&2008 Good Design Award

2006 Finalist of International Competition for Rainber "STERNBRAUEREI" in Salzburg (Austria)

2005 UIA Barbara Cappochin Architecture Prize (Italy)

Finalist of JIA Architectural Awards (Japan)

Good Design Award (Japan)

Awarded Architectural Digest 5 ans Exposition (France)

CINO ZUCCHI ARCHITETTI

The Cino Zucchi Architetti studio is constantly searching new spatial solutions for contemporary life in the delicate and rapidly changing context of the European landscape. Its goal is to combine innovation and research with professional completeness capable of responding to complex programs on any scale and employing, when needed, a well-established net of specialized consultants.

The studio has designed and realized many commercial, public, industrial and residential buildings, public spaces, renewal plans and master plans for agricultural, industrial and historical areas. It has participated in numerous national and international competitions and is active in the field of interior and exhibition design. The works of the studio have received widespread critical acclaim and have been published in magazines both in Italy and abroad, and have been the subject of numerous exhibitions.

KENGO KUMA

1954 Born in Yokohama, Kanagawa Prefecture, Japan

1979 Completed the Master Course, Department of Architecture, Graduate School of Engineering, The University of Tokyo

1985 – 1986 Visiting Scholar Graduate School, Columbia University and Asian Cultural Counci (USA)

1990 Established Kengo Kuma & Associates

1994 Lecturer at the Graduate School of Architecture, Planning and Preservation, Columbia University (New York, USA)

1998 – 1999 Professor at the Faculty of Environmental Information, Keio University

2001 – 2009 Professor at the Faculty of Science and Technology, Keio University

2007 – 2008 Visiting Professor at the School of Architecture, University of Illinois at Urbana-Champaign (Chicago, USA)

2008 Established Kuma & Associates Europe (Paris, France)

2009 – Present Professor at the Graduate School of Architecture, The University of Tokyo

TABANLIOĞLU ARCHITECTS

Consisting of more then 120 employees, operating basically in Turkey, Middle East and CIS Countries, Tabanlioglu has more then 50 years experience in architecture. Searching for new efficiencies in terms of global and environmental needs and developments, Tabanlıoğlu works comprise wide range of building types. Tabanlıoğlu has been awarded several international and national prizes: The company ranked 79th "BD World Architecture Top 100 list" in 2009. Partners Melkan Gürsel and Murat Tabanlıoğlu are selected as the Architect of the Year in 2010 by MEA- Middle East Architect Award. Recently, International Property Awards awarded Astana Arena the best public building in the Asia-Pacific and then in the World. Loft Gardens project of the office has received the International Award by Royal Institute of British Architects in 2011.

www.tabanlioglu.com

SOTIRIS TSOULOS

Sotiris Tsoulos was born in Athens Greece in 1974, has graduated from the School of architecture in Strathclyde University, Glasgow in 1998 and has worked for several years in Vienna, Athens, Kuala Lumpur and Dubai .

Tsoulos Associates Ltd was founded in 2005 in Athens and it has since developed to a team of 10 to 15 architects and interior designers with compass the unexpected solutions that architecture can provide.

The practice is collaborating extensively with large international based practices on large scale projects and deals with residential, interior, retail, hotel, public and any other architecture that exists.

New materials, new types of plans, economic viable approaches and human scale are the tools the office uses.

The way we achieve our goals is with extensive planning and site supervision and we try to follow the project all the way through to its last detail.

Tsoulos Associates doesn't believe that architecture can be bottled in categories but it's a way of thinking that captures the best of every project.

ABIN DESIGN STUDIO

ADS was founded in October 2005 by Abin Chaudhuri, an architecture graduate from Jadavpur University (J.U.) with a specialization in Industrial Design from Domus Academy, Milan. Jui Mallick, his classmate from J.U., joined him in early 2006 and went on to becoming his partner. What started off as a small 3 person firm is now a frontline organization rendering complete design and management solutions right from conceptualization to realization of the space, object or visual in the market, keeping in mind the synergy of execution of design solution through extensive research and innovation. We are into architecture, interior design (corporate and retail), design of exhibition spaces, industrial design and graphic design.

A-LAB

A-lab was founded in 2000 as a collaborative architecture and urbanism office, and it has today two partners, Geir Haaversen and Odd Klev.

Like most other young practises, in the early years we worked to establish a position and create our identity as an office.

From 2002 to 2005 a-lab won the competitions for the Oslo-Barcode Master plan (collaboration with DARK architects and MVRDV), the IT-Fornebu Office Building (Oslo) and the Arctic Culture Centre in Hammerfest. In 2009 the projects IT-Fornebu and Arctic Culture Centre were completed.

A-labs latest wins are the competitions for sustainable urban development for Furuset in Oslo and the Skien waterfront development. The office also won the competition for new border stations in Sør-Varanger in 2010.

The last years our concern and challenge has been to get our concepts and ideas built. Currently a-lab is building the DnB NOR Headquarters in collaboration with DARK and MVRDV, the Statoil Office Building in Oslo and the Vadsø Museum.

MARIO CUCINELLA

Mario Cucinella, 1960, founded Mario Cucinella Architects in Paris in 1992 and in Bologna in 1999. He received since the beginning of its activities, major awards in international competitions. Cucinella is Visiting Professor at Nottingham University and regularly lectures in Italy and abroad.

MC avail itself of a team of architects and engineers from various countries and has a solid background in architectural design. The firm is also active in industrial design and technological research – through partnerships with universities and research programs of the European Commission – to which it devotes a special internal unit.

He has received numerous awards, including the MIPIM Green Building Award (Cannes, 2011); the MIPIM Green Building Award (Cannes, 2009); the Energy Performance+Architecture Award (Paris, 2005), the Special Award for the Environment by the World Architecture Congress (Dubai, 2005), the Outstanding Architect 2004 by the World Renewable Energy Congress (Denver, USA); the Kunstpreis 99 – prize for architecture - from the Akademie der Künste (Berlin, 1999).

DOMINIQUE PERRAULT

Figure of French architecture, Dominique Perrault gained international recognition after having won the competition for the National French Library in 1989 at the age of 36. This project marked the starting point of many other public and private commissions abroad.

As his approach is to transform the landscape and territories, Dominique Perrault's work as an architect is strictly related to his role as an urban planner. His urban studies in France and abroad help describe the continuity and consistency of his method.

Perrault, member of the Grand Paris scientific council, was appointed curator of the French Pavilion in the 12th Architecture Biennale in Venice (2010), being the subject of the installation METROPOLIS.

Dominique Perrault received many prestigious prizes and awards, including the AFEX Award for the Ewha Womans University in Korea and the "Grande Médaille d'or d'Architecture" from the Académie d'Architecture in 2010, the Mies van der Rohe prize (1997), the French national Grand Prize for Architecture (1993) and the Equerre d'argent prize for the Hotel Industriel Berlier (1989).

CEBRA

CEBRA is a Danish architectural office with its background in the architectural circles of Aarhus. The office was founded 2001 by the architects Mikkel Frost, Carsten Primdahl and Kolja Nielsen – all graduates from The Aarhus School of Architecture.

In 2008 CEBRA received the most prestigious award in Scandinavia the "Nykredits Architecture Award" and in 2006 CEBRA received the Golden Lion at the Venice Biennale 2006 for the project Co-evolution, which has been exhibited at the Biennale, São Paolo, Beijing, Manchester, Copenhagen and Venice 2006/2007. In 2008 CEBRA participated in the Biennale once again – this time curating the Danish pavilion at the Venice Biennale ECOTOPEDIA - Walk the Talk. The Bakkegaard School was nominated for the Mies van der Rohe award in 2006.

The office has worked with numerous architectural functions and scales but dwellings, sports facilities and school buildings have been predominant.

CHO BYOUNG-SOO

REGISTRATION:

Licensed and Registered in the state of Montana, USA

Member of American Institute of Architects, AIA

Member of National Council of Architectural Registration Boards, NCARB

Registered in Seoul, Korea (KIA)

EDUCATION

Harvard University, Cambridge, Massachusetts, USA

1991 Master in Architecture (MARCH II)

1991 Master of Architecture in Urban Design ACTIVITY

1990 Attended Harvard-ETH Student Exchange Program, Zürich, Switzerland

ACTIVITY

1990 Traveling Scholarship to Turkey, Federal Government of Switzerland

Montana State University, Bozeman, Montana, USA

1986 Bachelor of Architecture

1986 Winner, New Sweden Student Urban Design International Competition

AWARDS

2010 Swoo Geun Kim Cultural Prize for Earth House

AIA Citation Award, Northwest and Pacific Regional for design of Earth House

2009 AIA Honor Award, Montana chapter for the design of Earth House

AIA Honor Award, Montana chapter for the design of Hanil Cement Information Center and Guest House

AIA Honor Award, Northwest and Pacific Regional for the design of D-Shaped Concrete Box House

KIA National Award, for the design of Hanil Cement Information Center and Guest House

KIA National Award, (Areomsol Kindergarten), Korea

ANDRÁS VIKÁR DLA & ISTVÁN LUKÁCS

After fifteen years experience of architectural design and city planning in the design office of the city council of Budapest, Hungary, István Lukács and András Vikár established their own architecture office. The company was constantly growing in the last decade, total building volume is over 500,000 sqm. The firm is engaged in Architectural design, from concept to detailed planning, open for new challenges. Thorough work and high-quality building materials are extremely important in their design. The studio's main principle is to create contextual, well functioning, sustainable architecture based on clean geometries.

PRIZES AND AWARDS

1996	Award of the city of Budapest, (office building Városmajor utca, Budapest)
2000	Observer architectural prize (Budapest, Jagellóút)
2001	Pro Architectura prize (Graphisoft building, Budapest)
2008	Ybl Architectural Award
2009	Pro Architectura prize (Graphisoft Park Microsoft building Budapest)
2010	Újbuda Pro Architectura prize (Simplon B)
2011	Europen Property Award (Akadémia Park Officium) CEE Green Building Awards (Akadémia Park Officium) CIJ Awards Hungary (Akadémia Park Officium)

HELLER MANUS ARCHITECTS

Heller Manus Architects is an internationally recognized, San Francisco-based architecture and urban design firm, with offices in San Francisco and Shanghai. Since founding the firm in 1984, Jeffrey Heller, FAIA, the firm's president and councilor of the Copenhagen-based, international Green Growth Leaders, and Clark Manus, FAIA, the firm's CEO and 87th President of the American Institute of Architects, have dedicated more than 25 years to develop a diversified, client-oriented firm that is a design and sustainability leader in the profession.

Heller Manus Architects' depth of experience encompasses a wide range of new construction and historical renovation projects including residential, mixed-use, commercial, master plans and urban design, hotel, retail, academic, civic, rehabilitation and adaptive re-use, entertainment, and transportation. Since 2004, the firm's practice area has grown to include several major projects throughout China including Shanghai, Guangzhou, Inner Mongolia, and the greater Beijing area. The firm is a leader in the profession and green movement with some of the first sustainable architecture and large scale urban planning projects in the US and China.

www.hellermanus.com

SCHULTE-FROHLINDE ARCHITEKTEN

In 1997, SFA was founded by Mr. Dipl.Ing architect HSF. With years of experience in the renovation of old-buildings and the implementation of conservation projects and since 2003 with new buildings and major projects, the office employs today an interdisciplinary and international team of 10 to 15 employees. Young creative architects aside experienced engineers work on task definition to the project realization with a focus on sustainable design and the latest energy concepts.

Sandra Pauquet since 2004 freelancer, still collaborating with the agency and the parisian stylist Isabelle Baudry, and was charged of the exclusive interior design of the SOLON headquarter.

BEHNISCH ARCHITEKTEN

Behnisch Architekten was founded in 1989 and works out of three offices – Stuttgart, Munich, and Boston. These offices are directed by Stefan Behnisch and his partners Robert Hösle, Robert Matthew Noblett and Stefan Rappold in varying combinations.

From the outset, the social dimension of architecture has been a fundamental aspect of the firm's design philosophy. The search for innovative and sustainable solutions making optimum use of natural resources has produced a rich variety of buildings, each of which responds to specific user requirements and site conditions.

Behnisch Architekten realized such signature projects as the Institute for Forestry and Nature Research in Wageningen, the Netherlands (1998); the LEED Platinum Genzyme Center in Cambridge, MA (2004); as well as the new Unilever Headquarters (2009) and the Marco Polo Tower in Hamburg's Hafen City (2010). The most recently completed major construction project is a new administrative building for the World Intellectual Property Organization in Geneva (2011) which is adjacent to the Organization's main headquarters building. Over the years the practice has established an international reputation as a firm that combines design excellence with advanced expertise in sustainability, and the firm's innovative work has been awarded widely.

ACXT

ACXT, an architecture services provider, started off as a society of professionals who came together with the aim of offering a quality alternative within the field of architecture, becoming a valid option for the professional development of architects interested in high level architecture. Its philosophy marks the creation of a professional framework of collaboration and avail of its members' synergies, always from a full understanding of architecture: both technically and culturally.

The philosophy that guides ACXT is based on the respect for traditional values, still in force today, generated during the secular practice of architecture. ACXT is committed to signature architecture, understood in terms of respect for the creative values that architects, with a clear customer orientated practice, have historically contributed throughout the whole intellectual and constructive process. ACXT aims to offer a quality solution to the huge social credit that the practice of architecture implies in its double commercial and cultural dimension.

ACXT-IDOM has become the largest provider of professional services to architecture in Spain and one of the most relevant worldwide. In the January 2012 edition of the BD World Architecture's Annual Survey, a British publication that gathers the hundred most important architecture firms worldwide, ACXT-IDOM was ranked number 48.

www.acxt.net

MARIO MAZZER

Born in 1955, he got the Architecture Degree at the Milan Polytechnic in 1978 and the diploma in Industrial Design at the Polytechnic School of Design in 1979. He gained experience in the 1970s working with masters of Italian design, like Achille Castiglioni and Marco Zanuso. In 1980 opened his studio at Conegliano Veneto (Italy).

He received special mentions in 1988 to 1989 at the Young & Design competition, winning it in 1991. He was selected in 1991 for the Design et Mutations exhibition in Paris, in Genova for the Tecnhotel trade fair, and in 1996 with various products for the event I Progettisti del Triveneto. In 1996 he received an honorable mention from the ID Annual Design Review New York. In the last years he has taken part in the itinerant exhibition Segnali di cibo and its related television programs. He won the International Grandesign award for the jeweller's category and was mentioned in the 5th edition. Many of his products were selected for the ADI Design Index and International Design Yearbook. In 2001 his monography *Thinking about Things: Mario Mazzer designer* was published. In 2006 and 2007 he received a special mention for the Riabita Competition. In 2009 receives the second prize for the desk "OLA" designed. In 2009, participates in the 3rd exhibition of Linking People at Abitare il Tempo. In 2009 wins the "GOOD DESIGN Award" of Chicago. In 2010 wins the 2010 Red Dot Design with all three participating products.

AUER+WEBER+ASSOZIIERTE

Auer+Weber+Assoziierte
Stuttgart | Munich, Germany

The architectural office Auer+Weber+Assoziierte originates from the Auer+Weber architectural group existing since 1980. With offices in Munich and Stuttgart, the company employs an average of between 90 and 100 personnel.

The focal point of activities is the conception and design of new buildings and alterations for public-sector institutions and private clients, increasingly as total service planning consultants or in collaboration with investors and construction companies. Over the past few years, we have been increasingly involved in the design and realization of international projects.

The common characteristic of all projects lies in the distinctive and coherent development of architectural which emerges from the specific brief and unique site conditions.

EHRLICH ARCHITECTS

Ehrlich Architects is internationally recognized for distinctive design, which extends the traditions of California modernism through an approach that fuses new technologies with cultural and environmental sensitivity. The firm has been recognized with prestigious accolades including eight National American Institute of Architects (AIA) awards.

Embracing the convergence of complex global factors that challenge today's architects, the firm has evolved a unique approach to architecture and planning called Multicultural Modernism. Largely developed through the travels and experiences of founding Principal Steven Ehrlich, Multicultural Modernism centers around architectural anthropology and incorporates four key elements: sensing place and listening to people; courtyards as an antidote to density and stress; the influence of Los Angeles, an "incubator of change"; and cross-cultural fusion.

Each of the firm's commissions is designed with a straightforward approach to form and function that maximizes environmentally friendly materials and building methods.

www.ehrlicharchitects.com

ENRIC BATLLE I DURANY AND JOAN ROIG I DURÁN

They were born in Barcelona in 1956 and 1954, respectively. They studied at the Barcelona School of Architecture (ETSAB) and qualified as architects in 1981. In the same year, they set up BATLLE I ROIG, ARQUITECTES, and won the competition for the Roques Blanques cemetery in El Papiol, Barcelona, which became the practice's first built work. Since then, they have produced many projects in the fields of architecture, landscape architecture and urban planning.

They have finished numerous works and also have been awarded numerous prizes.

Enric Batlle lectures in Urbanism and Landscape Architecture at the Vallès School of Architecture (ETSAV) and on the Master's degree in Landscape Architecture at the Polytechnic University of Catalonia (UPC).

Joan Roig has taught Architectural Projects at the ETSAB since 1984, and has lectured at the Barcelona School of Agriculture and been visiting professor at many schools.

VAILLO+IRIGARAY

The Architecture Studio called Vaillo+Irigaray (Architectural Strategies, SAS S.L.) was founded in September 1997, becoming in a limited partnership the 21st of March 2000, with the Notary of Pamplona called D. Francisco Salinas Frauca and registered in the Commercial Register of Navarre, at Volume 741, Folio 31, Paper Number NA-15244, first registration, on the 10th of April 2000.

On the 3rd of June 2010, VAILLO-IRIGARAY and ASOCIATES S.L.P. association is founded because of the addition of new partners, established in PAMPLONA, TAFALLA Road, 31 Low, by public deed given by the Notary of PAMPLONA called D. ERNESTO RODRIGO CATALÁN, following the 1173/10 Protocol and registered in the Commercial Register of Navarre dated on the 15th of September 2010, at Volume 1532, Folio 195, Paper Number NA-30455, registration number 000001.

FERDINAND AND FERDINAND ARCHITECTS

The architectural office was established by the three of us in 1994. The company name was Secrys Építészeti Tervező és Szolgáltató Bt. at that time. It switched to Secrys Építészeti Tervező Kft. (Secrys Architectural Design, Inc.) in 1996 (owners: Ágnes Sebestyén, Csaba Ferdinánd, Árpád Ferdinánd) as to follow the growth of the firm. To emphasize the importance of their names they established a new company in 1998, named Ferdinánd és Ferdinánd Építésziroda Kft. (Ferdinand and Ferdinand Architects, Inc., Partners: Csaba Ferdinánd, Árpád Ferdinánd), however they have kept on fulfilling their previous commitments in the frame of our previous firm.

The office is equipped with ten personal computers, four professional printers and a plotter, they utilize a 100 Mbytes/sec internal network to share their projects. These features help them to keep high standards in their work, and delivering strictly on schedule. Their work is acknowledged by the Treasury Property Departement and well-established private investors among their clients.

"Their goal is the clear, modern architecture. They are looking for the possibile in unpossible. They know that they are only little points in the all and the world will getting better but they are always beliving in it. That's way because they are optimistic. The song is important not the singer."

SKYLINE ARCHITEKTEN

Established seven and a half years ago, skyline architekten ZT GmbH has released a publication of an overview of projects completed since 2003.

Our range of references includes:

High-rise buildings – office construction

Hotel and spa facilities

Hospital construction

Industrial buildings

Residential buildings

Special purpose buildings

Our professional competence also covers expert opinions and project management:

Hospital construction

Fire prevention (fire protection planning)

BDP

BDP is the foremost interdisciplinary practice of architects, designers, engineers and urbanists in Europe.

They work closely with users, clients and the community to create special places for living, working, shopping, culture and learning across Europe, Africa, Asia and Australia.

Founded in 1961, they now employ over 900 architects, designers, engineers, urbanists, sustainability experts, lighting designers and acoustics specialists in studios across the UK, Ireland, Netherlands, UAE, India, and China. BDP has a leading track record in all major sectors including health, education, workplace, retail, urbanism, heritage, housing, transport and leisure.

They combine expertise across disciplines, locations, sectors and all major building types to deliver a truly integrated way of working – resulting in high quality, effective and inspiring built spaces.

SABBAGH ARQUITECTOS

The office was founded in 1986, by the architect Juan Sabbagh P., and the designer Mariana Sabbagh P. Both graduated from University of Chile.

In the year 2000 the office was join by Juan Pedro Sabbagh architect graduated from Catholic University and in the 2003 the architect Felipe Sabbagh graduated from Universidad de Chile joins the office too.

In 2002 Juan Sabbagh P. gets the national prize of architecture and he is a professor in a workshop at the Faculty of Architecture of Universidad de Chile since 1994 to 2004. Between 2004 to 2006 was the president architecture college of Chile.

The office have developed his professional works in many areas specialized in industries, services, education and corporative buildings. It has a distinguish participation at the Chilean architecture Festivals with many distinctions in different projects.

DAMILANO STUDIO

The founder of the studio is Duilio Damilano, his passion for volumes, instead, comes from a family of sculptors. His father and his brother have, in fact, passed on to him an interest for the plastic and material aspect of every sculpture or architecture. Duilio Damilano says that he's always been attracted by architectures since childhood.

His design path begins from the study of the light and how this affects and moulds the shapes. "It is an instinctive vocation – says Duilio Damilano – a feeling that starts from deep inside and only at a later time is conveyed inside the channels of rationality. Certainly is something that's always been part of my approach to architecture and which goes into my entire project. I mean the kind of lighting that goes beyond the physical aspect of the luminous element to become an integral part of the architectural design. A project can't peter out with the daylight."

Major areas of interest of Damilano Studio are the design of residential buildings, offices and commercial and receptive structures, both in Italy and abroad. An ongoing collaboration has occurred with the architects Claudia Allinio, Alberto Pascale, Enrico Massimino, Jessica Pignatta.

后记

本书的编写离不开各位设计师和摄影师的帮助，正是有了他们专业而负责的工作态度，才有了本书的顺利出版。参与本书的编写人员有：

Abin Chaudhuri, Adnan Harambasic, Aitor Ortiz, Alessandro Vergot, Ali Eray, Amaia Lastra, Amral Imran , András Klopp, András Vikár, Andrea Sartori, Angel Gómez, Angelo Agostini, Arrate Achalandobaso, Asplan Viak, Balázs Domokos, Bernardo Fort Brescia, Laurinda Spear, Çağrı Akay, Caterina Maciocco, Cemal Emden, Cho Byoung-soo, Cino Zucchi, Civertan, Claudia Allinio, Cristina Lamikiz, Daniel Weinbach, Daniele Domenicali, David Hirsch, Debora Venturi, Derek Heslop, Dominique Perrault, Duilio Damilano, Electro, Elizabeth Francis, Emanuele Meinero, Enric Batlle, Joan Roig, Enrico Casetta, Alberto Pascale, Eric Druse, Erik Horn , Eva Cantwell, Eva Serrats, Fancisco Donadio, Fernado Ortega, Flack + Kurtz, Francesca Castagna, Francesco Fulvi, Francisco García, Gábor F. Tóth, Geir Haaversen, George Sexton, Gerardo Rodríguez, Gianluca Moras, Giuseppe Perrone, Goretti Guillén, Guo Xiaohui, Helena Salvadó, Heller Manus, Ibai Ormaza, Igor Udaondo, Inger Totland, István Lukács, Jan Petter Seim, Jarand Midtgaard, Javier Pérez Uribarri, Jeffrey Heller, Jessica Pignatta, Jose Cavallero, Jui Mallik, Jun Murata, Kang Woo-hyun, Kaushik Majumder, Kengo kuma, Krueck & Sexton, Kwon Sung-Hwa, Laurinda Spear, Luca Donati, Luca Stramigioli,, Luis Fonseca, Manuela Taricco, Marc Rips, Marco Da Ros, Marco Mirisola, Marcus Bredt, Marina Duran, Mario Cucinella, Mario Mazzer, Melkan Gürsel, Murat Cengiz, Natalino Roveri, Nicholas Locke, Norman McGrath, Nuno Lobo, Odd Klev, Oscar Costa, Oscar Malo, Ozan Öztepe, Pai Yong-eun, Park Gi-hyun, Park Joo- hyun, Paul Maurer, PGI Grup SL Post-Tension system, SRC, Pradip Sen, Prakash Patel, Rafael Pérez, Rambøll Norge, Read Jones Christofferson Calgary, Richard Ceccanti, Roberto Aparicio, Roberto Fdez de Gamboa, Robin Hill, Rodolfo Cerati, Rogan Coles, Roland Halbe, Rupert Steiner, Russell Abraham, Seo Ji-young, Silke Reents, Sotiris Tsoulos, Soutaro Hayashi, Steel Frame, Stefania Neodo, Svigals, Szandra Váraljai, Takashi Yamaguchi, Tamás Bujnovszky, Tim Griffith, Tishman Speyer, Tonje Løvdahl, Tor Inge Hjemdal, Vivian Manasc,Xabier Aparicio Ortega,Ziortza Bardeci

ACKNOWLEDGEMENTS

We would like to thank everyone involved in the production of this book, especially all the artists, designers, architects and photographers for their kind permission to publish their works. We are also very grateful to many other people whose names do not appear on the credits but who provided assistance and support. We highly appreciate the contribution of images, ideas, and concepts and thank them for allowing their creativity to be shared with readers around the world.